"十三五"高等学校规划教材

计算机网络实验指导

陶 骏 主 编
伍 岳 副主编
周鸣争 颜云生 张云玲 参 编

中国铁道出版社有限公司
CHINA RAILWAY PUBLISHING HOUSE CO., LTD.

内 容 简 介

本书共包括 36 个计算机网络实验，内容丰富、理论联系实际、深入浅出、循序渐进、图文并茂；在强调基本理论的基础上，引入大量的实例来阐明各种技术与应用，力求做到知识性、实用性与综合性的有机结合。

本书适合作为普通高等学校计算机信息类专业的计算机网络基础、计算机网络管理等课程的实验教材，也可作为成人计算机网络的培训教材及计算机网络自学者的参考用书。

图书在版编目（CIP）数据

计算机网络实验指导/陶骏主编. —北京：中国铁道出版社，2018.8（2024.7 重印）
"十三五"高等学校规划教材
ISBN 978-7-113-24785-0

Ⅰ.①计… Ⅱ.①陶… Ⅲ.①计算机网络-实验-高等学校-教学参考资料 Ⅳ.①TP393-33

中国版本图书馆 CIP 数据核字(2018)第 171780 号

书　　名：	计算机网络实验指导
作　　者：	陶骏
策　　划：	翟玉峰　刘梦珂　　　　编辑部电话：（010）51873135
责任编辑：	翟玉峰　卢　笛
封面设计：	郑春鹏
责任校对：	张玉华
责任印制：	樊启鹏

出版发行：中国铁道出版社有限公司（100054，北京市西城区右安门西街 8 号）
网　　址：https://www.tdpress.com/51eds/
印　　刷：三河市宏盛印务有限公司
版　　次：2018 年 8 月第 1 版　2024 年 7 月第 6 次印刷
开　　本：787 mm×1092 mm　1/16　印张：9.75　字数：230 千
印　　数：5 801～7 000 册
书　　号：ISBN 978-7-113-24785-0
定　　价：28.00 元

版权所有　侵权必究

凡购买铁道版图书，如有印制质量问题，请与本社教材图书营销部联系调换。电话：（010）63550836
打击盗版举报电话：（010）63549461

Foreword 序

《计算机网络实验指导》是全日制大学本科计算机专业和相近专业的专业基础课"计算机网络"的配套教材,是学习计算机网络相关知识的重要课程。它系统介绍了计算机网络的基本理论、基本知识、基本技术、基本应用的内容,为后续学习与计算机网络相关的课程打下基础。通过学习,要求学生系统地了解计算机网络的基本原理,掌握计算机网络的基本概念和计算机网络在日常生活和工作中的应用方法,学会计算机网络的实际操作和日常管理及维护。

计算机网络通俗地说就是由多台计算机(或其他计算机网络设备)通过传输介质和软件物理(或逻辑)连接在一起组成的。总的来说,计算机网络的组成基本包括计算机、网络操作系统、传输介质及相应的应用软件四部分。

学习上述的计算机网络四个部分的一个重要的辅助手段就是进行计算机网络实验,通过认真系统地进行实验,有助于学生掌握计算机网络的基本知识,并能够在实践中灵活应用。本书包含了计算机网络体系中大部分的基础实验,脉络清晰,可操作性强。本书难能可贵的是在软件模拟器实验的基础上提供了基于H3C设备的网络实验,可以使读者在真实的网络环境下学习计算机网络知识。另外,本书的网络实验也涉及了目前比较常用网络协议,如ISIS和BGP。

相信通过对本书的内容学习,读者的计算机网络理论水平和实际操作水平都能更上一个台阶!

华三通信技术有限公司资深网络构架师 岑海洋

2018年7月

Preface 前　言

"计算机网络"是计算机与信息类相关专业一门重要的专业基础课程，也是计算机网络从业人员必须具备的基础知识。计算机网络分成拓扑和协议两部分，协议大都被分层设计方法屏蔽，不利于理解和掌握，本书的编写目的是让读者在拓扑设计的基础上进行计算机网络实验，并在实验中观察和分析网络协议，以增强工程动手能力与对计算机网络基本原理的理解。

本书根据计算机网络技术的发展与应用需求，在计算机网络的分层结构基础上，考虑其基础性、综合性与设计性的要求，设计了 36 个计算机网络实验，内容丰富、系统全面，实验编排科学合理，有较大的选择空间，可满足不同的教学要求。

实验涵盖了计算机网络体系中的各个层次：物理层、数据链路层、网络层、传输层、应用层。物理层包含双绞线的制作、无线 AP 使用等；数据链路层包含 VLAN 划分、VLAN 透传等；网络层包括静态路由协议、RIP、OSPF 协议的使用实验等；传输层包括 TCP 会话程序的编制等；应用层包括 FTP、DNS 等协议的使用实验，并在此基础上设计了一些规模较大的综合实验，如网吧网络的设计、校园网络设计等。

本书对大多数实验都设计了模拟器软件实验和真实设备实验，建议读者先进行模拟器软件实验，再进行真实设备实验，这样有助于掌握基本的网络原理和技术。实验后可选择做一下相关的思考题，这样更有助于掌握基本的网络原理和技术，从而在现实生活中能够灵活熟练地运用计算机网络知识。

本书由安徽信息工程学院陶骏任主编，伍岳任副主编。具体分工如下：实验 1～实验 8 由陶骏编写，实验 9～实验 16 由周鸣争编写，实验 17～实验 24 由伍岳编写，实验 25～实验 36 由颜云生编写，附录 A～C 由张云玲编写。陶骏与周鸣争负责全书内容的选材和统稿工作。

本书受安徽省教育厅自然科学重点科技项目"基于深度学习的 SDN 网络研究支撑（项目基金号：KJ2018A0626）"资助。书中的实验由严志兵、王淼、陈雷、陈帅和赵慧慧等同学协助完成，特向他们表示感谢。

由于编者水平有限，书中难免存在不足和遗漏之处，恳请广大读者和同行批评指正。作者的联系电子邮箱：taonian@126.com。

编　者
2018 年 5 月

目 录

实验 1　双绞线的制作 1
实验 2　PC 网络指令测试 4
实验 3　交换机基本操作（思科模拟器） 7
实验 4　交换机基本操作（上机） 9
实验 5　交换机 VLAN 透传设置（思科模拟器） 11
实验 6　交换机 VLAN 透传设置（上机） 15
实验 7　VTP 配置实验（思科模拟器） 19
实验 8　VTP 配置实验（上机） 23
实验 9　静态路由（思科模拟器） 26
实验 10　静态路由（上机） 30
实验 11　单臂路由器配置（思科模拟器） 33
实验 12　单臂路由器（上机　路由器） 36
实验 13　单臂路由器（上机　三层交换机） 39
实验 14　RIP 路由（思科模拟器） 42
实验 15　RIP 配置实验（上机） 46
实验 16　EIGRP 路由（思科模拟器） 49
实验 17　OSPF 路由配置实验（思科模拟器） 53
实验 18　OSPF 路由配置实验（上机） 56
实验 19　ISIS 路由配置实验（华为模拟器） 60
实验 20　ISIS 路由配置实验（上机） 64
实验 21　BGP 路由配置实验（思科模拟器） 68
实验 22　BGP 路由配置实验（上机） 72
实验 23　ACL 配置实验（思科模拟器） 75
实验 24　ACL 配置实验（上机） 80
实验 25　Telnet 配置实验（思科模拟器） 83

实验 26	IPv6 静态路由配置实验（上机）	85
实验 27	NAT 配置实验（思科模拟器）	88
实验 28	NAT 配置实验（上机）	92
实验 29	TCP socket 编程	95
实验 30	DNS 和 Web 配置实验（思科模拟器）	99
实验 31	FTP 配置实验（思科模拟器）	104
实验 32	路由器端口镜像配置（上机）	109
实验 33	无线网络设计（思科模拟器）	112
实验 34	无线网络设计（上机）	115
实验 35	网吧网络设计（思科模拟器）	117
实验 36	校园网络设计（思科模拟器）	122
附录 A	思科模拟器	129
附录 B	华为模拟器	131
附录 C	H3C 设备简介	137

实验 1 ➡ 双绞线的制作

背景介绍

双绞线（Twisted Pair，TP）是一种综合布线工程中最常用的传输介质，是由两根具有绝缘保护层的铜导线组成的。把两根绝缘的铜导线按一定密度互相绞在一起，每一根导线在传输中辐射出来的电波会被另一根线上发出的电波抵消，有效降低信号干扰的程度。

实验目的

掌握 RJ-45 双绞线网线的制作，掌握测试仪的使用方法。

实验步骤

（1）将双绞线剥皮如图 1-1 所示。

图 1-1 双绞线剥皮

（2）排线顺序，顺序为：橙白、橙、绿白、蓝、蓝白、绿、棕白、棕，如图 1-2 ~ 图 1-4 所示。

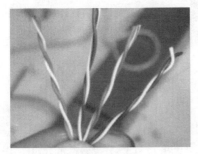

图 1-2 分线

图 1-3 排序

图 1-4 捋直

(3) 压进水晶头，如图 1-5 和图 1-6 所示。

图 1-5 修剪线头

图 1-6 压进水晶头

(4) 以同样的方式做网线的另一头。

实验验证

1. 物理检验

将网线插入测试仪，两个仪器 1~8 灯同时亮起，如图 1-7 所示，说明物理检验正常。

图 1-7 测试仪测试

2. 网络层检验

(1) 安装好串口线的驱动程序，准备好交换机登录软件 secure crt。

(2) 将做好的网线连在交换机任意两个业务口。此时交换机端口灯亮才表示正常，如图 1-8 所示。

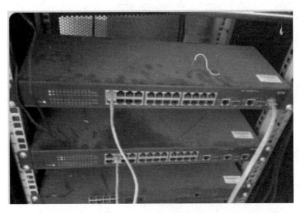

图 1-8　连接交换机

（3）分别登录两台交换机，给管理 VLAN 配置 IP 地址，两台交换机的 IP 地址在同一网段。

交换机 1 配置为：

```
sys
interface vlan-interface1
ip address 10.0.0.1 255.255.255.0
Quit
```

交换机 2 配置为：

```
sys
interface vlan-interface1
ip address 10.0.0.2 255.255.255.0
Quit
```

在交换机 2 上 ping 交换机 1 的管理 IP 地址，要是能 ping 通，就说明网络层也正常。

思考题

对于主机来说，交换机和路由器都属于异种设备，为什么交换机用直连线，而路由器用交叉线？

实验 ② PC 网络指令测试

背景介绍

　　ipconfig 实用程序和它的等价图形用户界面可用于显示当前的 TCP/IP 配置的设置值。这些信息一般用来检验人工配置的 TCP/IP 设置是否正确。除此之外，ipconfig 命令还有了解计算机当前 IP 地址、清除 DNS 缓存等作用，是进行测试和故障分析的必要项目。

　　ping（Packet Internet Groper，因特网探索器）是一种用于测试网络连接量的程序。它利用网络机器上 IP 地址的唯一性，给目标 IP 地址发送一个数据包，再要求对方返回一个同样大小的数据包来确定两台网络机器是否连通，延时多少。

　　tracert（Tracert Route，跟踪路由）用于确定 IP 数据包访问目标所采取的路径。tracert 先发送 TTL 为 1 的回应数据包，并在随后的每次发送过程中将 TTL 递增 1，直到目标响应或 TTL 达到最大值，从而确定路由。通过检查中间路由器发回的"ICMP 已超时"的消息确定路由。某些路由器不经询问直接丢弃 TTL 过期的数据包，这在 tracert 实用程序中看不到。

　　nslookup（Name Server Lookup，域名查询）可以指定查询的类型，可以查到 DNS 记录的生存时间，还可以指定使用哪个 DNS 服务器进行解释。在已安装 TCP/IP 协议的计算机上均可以使用这个命令，主要用来诊断域名系统基础结构的信息。

　　netstat 是控制台命令，是监控 TCP/IP 网络的非常有用的一个工具，它可以显示路由表、实际的网络连接，以及每一个网络接口设备的状态信息。netstat 用于显示与 IP、TCP、UDP 和 ICMP 协议相关的统计数据，一般用于检验本机各端口的网络连接情况。

实验目的

　　学会使用 ping, ipconfig, tracert, nslookup, netstat 等常见网络测试命令，进行检测网络连通、了解网络的配置状态，跟踪路由诊断域名系统等相关网络问题。

实验步骤

指令测试

（1）ipconfig 命令测试如图 2-1 所示。

　　因为连接的是 Wi-Fi，所以以太网的媒体状态为断开连接。除此之外，ipconfig 命令下还会显示所连接的 Wi-Fi、蓝牙、网卡等相关信息。

（2）ping 命令测试结果如图 2-2 所示。

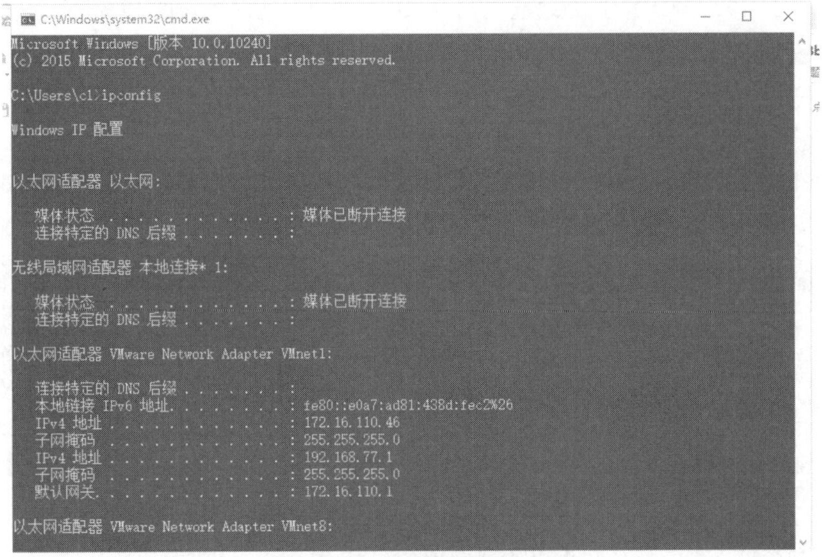

```
C:\Windows\system32\cmd.exe
Microsoft Windows [版本 10.0.10240]
(c) 2015 Microsoft Corporation. All rights reserved.

C:\Users\cl>ipconfig

Windows IP 配置


以太网适配器 以太网:

   媒体状态  . . . . . . . . . . . . : 媒体已断开连接
   连接特定的 DNS 后缀 . . . . . . . :

无线局域网适配器 本地连接* 1:

   媒体状态  . . . . . . . . . . . . : 媒体已断开连接
   连接特定的 DNS 后缀 . . . . . . . :

以太网适配器 VMware Network Adapter VMnet1:

   连接特定的 DNS 后缀 . . . . . . . :
   本地链接 IPv6 地址. . . . . . . . : fe80::e0a7:ad81:438d:fec2%26
   IPv4 地址 . . . . . . . . . . . . : 172.16.110.46
   子网掩码  . . . . . . . . . . . . : 255.255.255.0
   IPv4 地址 . . . . . . . . . . . . : 192.168.77.1
   子网掩码  . . . . . . . . . . . . : 255.255.255.0
   默认网关. . . . . . . . . . . . . : 172.16.110.1

以太网适配器 VMware Network Adapter VMnet8:
```

图 2-1　ipconfig 命令测试结果

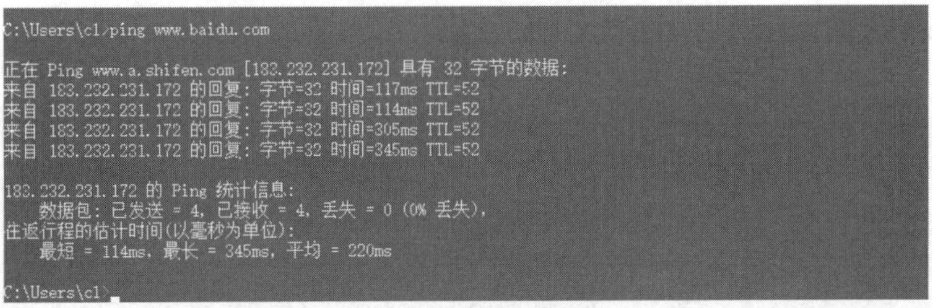

```
C:\Users\cl>ping www.baidu.com

正在 Ping www.a.shifen.com [183.232.231.172] 具有 32 字节的数据:
来自 183.232.231.172 的回复: 字节=32 时间=117ms TTL=52
来自 183.232.231.172 的回复: 字节=32 时间=114ms TTL=52
来自 183.232.231.172 的回复: 字节=32 时间=305ms TTL=52
来自 183.232.231.172 的回复: 字节=32 时间=345ms TTL=52

183.232.231.172 的 Ping 统计信息:
    数据包: 已发送 = 4，已接收 = 4，丢失 = 0 (0% 丢失)，
往返行程的估计时间(以毫秒为单位):
    最短 = 114ms，最长 = 345ms，平均 = 220ms

C:\Users\cl>
```

图 2-2　ping 命令测试结果

图 2-2 说明本机和 183.232.231.172 之间的网络很通畅。其中，“字节=32”表示本机对 183.232.231.172 发送的数据包是 32 字节，“时间=117 ms”表示数据从本机发送到 183.232.231.172 直到回到本机用时为 117 ms。“TTL=52”表示数据包在网络中存活的时间为 52 s。

（3）tracert 命令测试结果如图 2-3 所示。

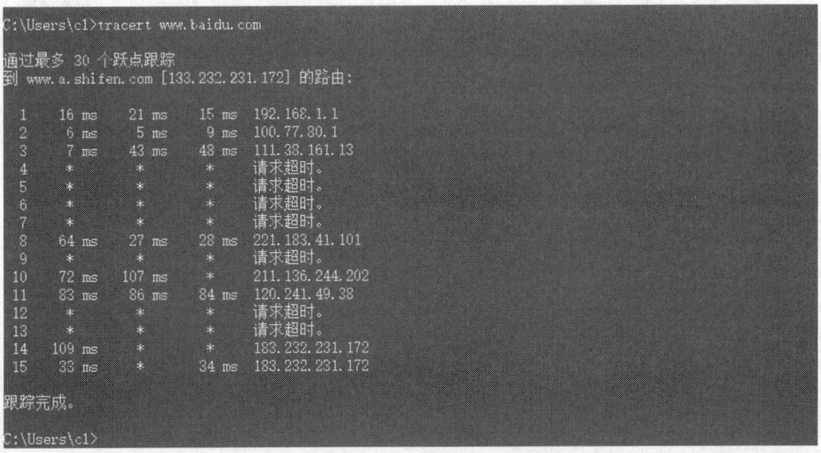

```
C:\Users\cl>tracert www.baidu.com

通过最多 30 个跃点跟踪
到 www.a.shifen.com [183.232.231.172] 的路由:

  1    16 ms    21 ms    15 ms  192.168.1.1
  2     6 ms     5 ms     9 ms  100.77.80.1
  3     7 ms    43 ms    48 ms  111.38.161.13
  4     *        *        *     请求超时。
  5     *        *        *     请求超时。
  6     *        *        *     请求超时。
  7     *        *        *     请求超时。
  8    64 ms    27 ms    28 ms  221.183.41.101
  9     *        *        *     请求超时。
 10    72 ms   107 ms     *     211.136.244.202
 11    83 ms    86 ms    84 ms  120.241.49.38
 12     *        *        *     请求超时。
 13     *        *        *     请求超时。
 14   109 ms     *        *     183.232.231.172
 15    33 ms     *       34 ms  183.232.231.172

跟踪完成。

C:\Users\cl>
```

图 2-3　tracert 命令测试结果

图 2-3 中的 192.168.1.1 等 IP 地址表示 ICMP 数据包到达的结点的 IP 地址，*表示所在的结点没有数据返回，即这个结点禁止 tracert 命令。

（4）nslookup 命令测试结果如图 2-4 所示。

```
Microsoft Windows [版本 10.0.10240]
(c) 2015 Microsoft Corporation. All rights reserved.

C:\Users\cl>nslookup
默认服务器:  AHHF-PS-SV35-DNS-HSL-Cache8
Address:  211.138.180.2

>
>
> www.baidu.com
服务器:  AHHF-PS-SV35-DNS-HSL-Cache8
Address:  211.138.180.2

非权威应答:
名称:    www.a.shifen.com
Addresses:  183.232.231.172
          183.232.231.173
Aliases:  www.baidu.com
```

图 2-4　nslookup 命令测试结果

输入该命令可以查看域名解析服务器的域名和 IP 地址。

（5）netstat 命令测试结果如图 2-5 所示。

```
C:\Users\cl>netstat

活动连接

  协议  本地地址          外部地址        状态
  TCP    127.0.0.1:50591        DESKTOP-OEMLF4L:50592  ESTABLISHED
  TCP    127.0.0.1:50592        DESKTOP-OEMLF4L:50591  ESTABLISHED
  TCP    127.0.0.1:51576        DESKTOP-OEMLF4L:51577  ESTABLISHED
  TCP    127.0.0.1:51577        DESKTOP-OEMLF4L:51576  ESTABLISHED
  TCP    127.0.0.1:64727        DESKTOP-OEMLF4L:64728  ESTABLISHED
  TCP    127.0.0.1:64728        DESKTOP-OEMLF4L:64727  ESTABLISHED
  TCP    127.0.0.1:64733        DESKTOP-OEMLF4L:64734  ESTABLISHED
  TCP    127.0.0.1:64734        DESKTOP-OEMLF4L:64733  ESTABLISHED
  TCP    127.0.0.1:64767        DESKTOP-OEMLF4L:64768  ESTABLISHED
  TCP    127.0.0.1:64768        DESKTOP-OEMLF4L:64767  ESTABLISHED
  TCP    192.168.1.100:50411    .:https                ESTABLISHED
  TCP    192.168.1.100:51557    180.97.249.163:http    CLOSE_WAIT
  TCP    192.168.1.100:51559    120.92.58.186:9002     CLOSE_WAIT
  TCP    192.168.1.100:51575    120.92.91.106:http     CLOSE_WAIT
  TCP    192.168.1.100:51578    120.92.96.2:5222       ESTABLISHED
  TCP    192.168.1.100:51581    120.92.91.106:http     CLOSE_WAIT
  TCP    192.168.1.100:51583    120.92.91.106:http     CLOSE_WAIT
  TCP    192.168.1.100:51649    183.232.96.107:https   CLOSE_WAIT
  TCP    192.168.1.100:51930    183.232.96.107:https   CLOSE_WAIT
  TCP    192.168.1.100:52144    183.232.96.107:https   CLOSE_WAIT
  TCP    192.168.1.100:52617    183.232.96.107:https   CLOSE_WAIT
```

图 2-5　netstat 命令测试结果

图 2-5 中所示为本机中一些网络接口的设备状态信息。

思考题

ping 本机的 IP 地址，再 ping 同组的一台计算机，要求指定数据包个数为 15 个，大小为 200 B，写出命令。

实验 ③ ➡ 交换机基本操作(思科模拟器)

背景介绍

双工模式分为半双工和全双工。半双工传输模式下双方不能同时通信,其通信率低,且有可能产生冲突,由于目前绝大多数都为交换网络,这种传输模式已经很少。全双工传输模式下双方则可以同时通信。

实验目的

掌握交换机的基本配置,理解交换机中双工模式、速率、自协商的含义。

实验步骤

1. 网络拓扑设计

实验所需要的设备为一台 PC,一台 2950-24 交换机。网络拓扑设计如图 3-1 所示。

2. 网络配置

交换机配置为:
```
//进入特权查看模式
enable
//进入配置模式
conf terminal
//设置交换机名称
hostname SW1
//进入端口 fastethernet0/1 配置模式
interface fastethernet0/1
duplex full(指定端口为全双工模式)
speed 100(指定端口速率)
```
注意:duplex auto 为自协商,duplex half 为半双工模式,本实验中设置的是全双工模式。

实验验证

网络验证如图 3-2 所示。

图 3-1 网络拓扑设计

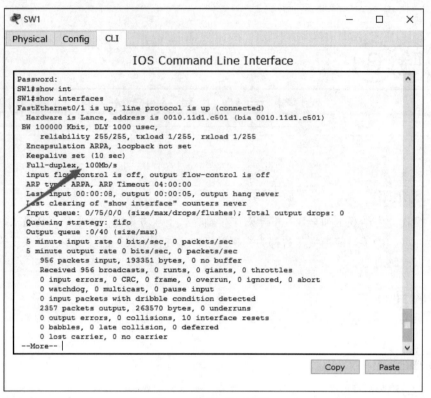

图 3-2　网络验证结果

怎样给交换机设置一个进入配置模式的密码？

实验 ④ 交换机基本操作（上机）

背景介绍
双工模式分为半双工和全双工。半双工传输模式下双方不能同时通信，半双工传输模式通信率低，且有可能产生冲突，由于目前绝大多数都为交换网络，这种传输模式已经很少。全双工传输模式下双方则可以同时通信。

实验目的
掌握交换机的基本配置，理解交换机中双工模式、速率、自协商的含义。

实验步骤

1. 网络拓扑设计
实验需要 H3C s3100v2-26TP-EI 交换机一台。网络拓扑设计如图 4-1 所示。

图 4-1　网络拓扑设计

2. 网络配置
交换机 SW1 的配置为：
```
//进入配置模式
sys
//命名
sysname SW1
quit
//配置端口
```

```
interface eth1/0/1
speed 10 (速率)
duplex full (全双工)
end
```

dis cu 命令下 1/0/1 端口属性如图 4-2 所示。

```
 ip address dhcp-alloc client-identifier mac vlan-interface1
#
interface Ethernet1/0/1
 description to pc
 port access vlan 10
 speed 100
 duplex full
#
interface Ethernet1/0/2
```

图 4-2 1/0/1 端口属性

怎样给交换机划分 VLAN？

实验 5 ➡ 交换机 VLAN 透传设置（思科模拟器）

背景介绍

透传指的是透明传输，即传输过程中，网络两端的设备感知不到传输存在。
VLAN 透传即数据帧结构不变，仍然带有 VLAN ID 信息往下传。

实验目的

掌握 VLAN 透传的原理及透传实验的基本配置方法。本实验通过思科模拟器实现。

实验步骤

1. 网络拓扑设计

实验需要两台思科 2960 交换机和四台 PC。网络拓扑设计如图 5-1 所示。

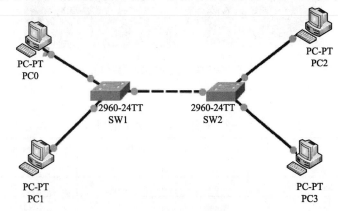

图 5-1 网络拓扑设计

2. 网络和计算机配置

四台 PC 的 IP 地址分配如表 5-1 所示。

表 5-1 四台 PC 的 IP 地址分配

设　备	接　口	IP 地 址	掩　码	网　关
PC0	Fa0	192.168.0.10	255.255.255.0	无
PC1	Fa0	192.168.0.100	255.255.255.0	无
PC2	Fa0	192.168.0.20	255.255.255.0	无
PC3	Fa0	192.168.0.200	255.255.255.0	无

在两台交换机未做任何配置的情况下,两台交换机就是两台集线器,其把四台计算机在物理上进行了连接并组成了一个局域网,此时计算机 PC0、PC1、PC2 和 PC3 是可以相互访问的,相互可以 ping 通,在 PC0 上 ping 另外三台计算机的测试结果如图 5-2 所示。

```
PC>ping 192.168.0.20

Ping 192.168.0.20: 32 data bytes, Press Ctrl_C to break
From 192.168.0.20: bytes=32 seq=1 ttl=128 time=46 ms
From 192.168.0.20: bytes=32 seq=2 ttl=128 time=31 ms

--- 192.168.0.20 ping statistics ---
  2 packet(s) transmitted
  2 packet(s) received
  0.00% packet loss
  round-trip min/avg/max = 31/38/46 ms

PC>ping 192.168.0.100

Ping 192.168.0.100: 32 data bytes, Press Ctrl_C to break
From 192.168.0.100: bytes=32 seq=1 ttl=128 time=32 ms
From 192.168.0.100: bytes=32 seq=2 ttl=128 time<1 ms
From 192.168.0.100: bytes=32 seq=3 ttl=128 time=15 ms

--- 192.168.0.100 ping statistics ---
  3 packet(s) transmitted
  3 packet(s) received
  0.00% packet loss
  round-trip min/avg/max = 0/15/32 ms

PC>ping 192.168.0.200

Ping 192.168.0.200: 32 data bytes, Press Ctrl_C to break
From 192.168.0.200: bytes=32 seq=1 ttl=128 time=16 ms
From 192.168.0.200: bytes=32 seq=2 ttl=128 time=31 ms
From 192.168.0.200: bytes=32 seq=3 ttl=128 time=63 ms
From 192.168.0.200: bytes=32 seq=4 ttl=128 time=31 ms
From 192.168.0.200: bytes=32 seq=5 ttl=128 time=31 ms

--- 192.168.0.200 ping statistics ---
  5 packet(s) transmitted
```

图 5-2 交换机配置前 ping 测试结果

两台交换机的端口规划如表 5-2 所示。

表 5-2 两台交换机的端口规划

设备	端口	端口类型	所属 VLAN	对端设备	对端设备端口
SW1	Fa0/2	access	100	PC0	Fa0
	Fa0/3	access	200	PC1	Fa0
	Fa0/1	trunk	无	SW2	Fa0/1
SW2	Fa0/2	access	100	PC2	Fa0
	Fa0/3	access	200	PC3	Fa0
	Fa0/1	trunk	无	SW1	Fa0/1

交换机 SW1 的配置为:

```
//进入特权模式
enable
//进入配置模式
configure terminal
//划分 vlan
vlan 100
vlan 200
exit
```

```
//配置端口
interface fa0/2
no shutdown
switchport mode access
switchport access vlan 100
interface fa0/3
no shutdown
switchport mode access
switchport access vlan 200
interface fa0/1
no shutdown
switchport mode trunk
//端口描述
des to-sw-2 Fa0/1
```

交换机 SW2 的配置为：

```
//进入特权模式
enable
//进入配置模式
configure terminal
//划分vlan
vlan 100
vlan 200
exit
//配置端口
interface fa0/2
no shutdown
switchport mode access
switchport access vlan 100
interface fa0/3
no shutdown
switchport mode access
switchport access vlan 200
interface fa0/1
no shutdown
switchport mode trunk
//端口描述
des to-sw-1 fa0/1
```

交换机配置完毕后，把 PC0 和 PC2 划归了 VLAN 100，把 PC1 和 PC3 划归了 VLAN 200，把原来的一个局域网划分成两个局域网。

实验验证

此时 PC0 只能 ping 通同一个局域网里的 PC2，其无法访问 PC1 和 PC3，具体测试如图 5-3 所示。

```
PC>ping 192.168.0.20

Ping 192.168.0.20: 32 data bytes, Press Ctrl_C to break
From 192.168.0.20: bytes=32 seq=1 ttl=128 time=47 ms
From 192.168.0.20: bytes=32 seq=2 ttl=128 time=15 ms
From 192.168.0.20: bytes=32 seq=3 ttl=128 time=32 ms
From 192.168.0.20: bytes=32 seq=4 ttl=128 time=31 ms
From 192.168.0.20: bytes=32 seq=5 ttl=128 time=31 ms

--- 192.168.0.20 ping statistics ---
 5 packet(s) transmitted
 5 packet(s) received
 0.00% packet loss
 round-trip min/avg/max = 15/31/47 ms

PC>ping 192.168.0.100

Ping 192.168.0.100: 32 data bytes, Press Ctrl_C to break
From 192.168.0.10: Destination host unreachable
From 192.168.0.10: Destination host unreachable
From 192.168.0.10: Destination host unreachable
From 192.168.0.10: Destination host unreachable

PC>ping 192.168.0.200

Ping 192.168.0.200: 32 data bytes, Press Ctrl_C to break
From 192.168.0.10: Destination host unreachable
From 192.168.0.10: Destination host unreachable
From 192.168.0.10: Destination host unreachable
From 192.168.0.10: Destination host unreachable
From 192.168.0.10: Destination host unreachable
```

图 5-3 交换机配置后的 ping 测试结果

思考题

如果 PC4 加入 VLAN 100，如图 5-4 所示，则交换机 SW1 如何修改配置？

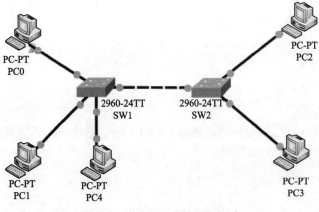

图 5-4 思考题的网络拓扑设计

实验 6 交换机 VLAN 透传设置（上机）

背景介绍

透传指的是透明传输，即传输过程中，网络两端的设备感知不到传输网络的存在。
VLAN 透传即数据帧结构不变，仍然带有 VLAN ID 信息往下传。

实验目的

掌握 VLAN 透传的原理及透传实验的基本配置方法。本实验通过华为 H3C 设备实现。

实验步骤

1. 网络拓扑设计

实验需要两台 H3C 交换机（s3100v2-26TP-EI 或者 s3600v2-28TP-EI 均满足）和四台 PC。网络拓扑设计如图 6-1 所示。

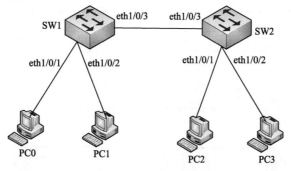

图 6-1 网络拓扑设计

2. 网络和计算机配置

四台 PC 的 IP 地址分配如表 6-1 所示。

表 6-1 四台 PC 的 IP 地址分配

设备	接口	IP 地址	掩码	网关
PC0	Fa0	192.168.0.10	255.255.255.0	无
PC1	Fa0	192.168.0.100	255.255.255.0	无
PC2	Fa0	192.168.0.20	255.255.255.0	无
PC3	Fa0	192.168.0.200	255.255.255.0	无

在两台交换机未做任何配置的情况下，两台交换机就是两台集线器，其把四台计算机在物

理上进行了连接并组成了一个局域网，此时计算机 PC0、PC1、PC2 和 PC3 是可以相互访问的，相互可以 ping 通，在 PC0 上 ping 另外三台计算机的测试结果如图 6-2 所示。

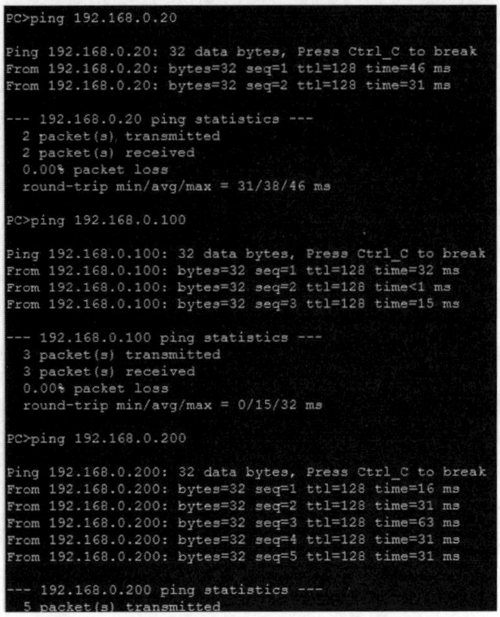

```
PC>ping 192.168.0.20

Ping 192.168.0.20: 32 data bytes, Press Ctrl_C to break
From 192.168.0.20: bytes=32 seq=1 ttl=128 time=46 ms
From 192.168.0.20: bytes=32 seq=2 ttl=128 time=31 ms

--- 192.168.0.20 ping statistics ---
  2 packet(s) transmitted
  2 packet(s) received
  0.00% packet loss
  round-trip min/avg/max = 31/38/46 ms

PC>ping 192.168.0.100

Ping 192.168.0.100: 32 data bytes, Press Ctrl_C to break
From 192.168.0.100: bytes=32 seq=1 ttl=128 time=32 ms
From 192.168.0.100: bytes=32 seq=2 ttl=128 time=1 ms
From 192.168.0.100: bytes=32 seq=3 ttl=128 time=15 ms

--- 192.168.0.100 ping statistics ---
  3 packet(s) transmitted
  3 packet(s) received
  0.00% packet loss
  round-trip min/avg/max = 0/15/32 ms

PC>ping 192.168.0.200

Ping 192.168.0.200: 32 data bytes, Press Ctrl_C to break
From 192.168.0.200: bytes=32 seq=1 ttl=128 time=16 ms
From 192.168.0.200: bytes=32 seq=2 ttl=128 time=31 ms
From 192.168.0.200: bytes=32 seq=3 ttl=128 time=63 ms
From 192.168.0.200: bytes=32 seq=4 ttl=128 time=31 ms
From 192.168.0.200: bytes=32 seq=5 ttl=128 time=31 ms

--- 192.168.0.200 ping statistics ---
  5 packet(s) transmitted
```

图 6-2 交换机配置前 ping 测试结果

两台交换机的端口规划如表 6-2 所示。

表 6-2 两台交换机的端口规划

设　备	端　口	端口类型	所属 VLAN	对端设备	对端设备端口
SW1	eth1/0/1	access	100	PC0	Fa0
	eth1/0/2	access	200	PC1	Fa0
	eth1/0/3	trunk	无	SW2	eth1/0/3
SW2	eth1/0/1	access	100	PC2	Fa0
	eth1/0/2	access	200	PC3	Fa0
	eth1/0/3	trunk	无	SW1	eth1/0/3

交换机 SW1 的配置为：

```
//进入配置模式
sys
//划分vlan
vlan 100
vlan 200
quit
//配置端口
interface eth1/0/1
port link-type access
port access vlan 100
interface eth1/0/2
port link-type access
port access vlan 200
interface eth1/0/3
```

```
port link-type trunk
port trunk permit vlan all
//端口描述
des to-sw-2 eth1/0/3
```

交换机 SW2 的配置为：

```
//进入配置模式
sys
//划分vlan
vlan 100
vlan 200
quit
//配置端口
interface eth1/0/1
port link-type access
port access vlan 100
interface eth1/0/2
port link-type access
port access vlan 200
interface eth1/0/3
port link-type trunk
port trunk permit vlan all
//端口描述
des to-sw-1 eth1/0/3
```

交换机配置完毕后，把 PC0 和 PC2 划归了 VLAN 100，把 PC1 和 PC3 划归了 VLAN 200，把原来的一个局域网划分成两个局域网。

实验验证

此时 PC0 只能 ping 通同一个局域网里的 PC2，其无法访问 PC1 和 PC3，具体测试如图 6-3 所示。

图 6-3　交换机配置后的 ping 测试结果

思考题

如果 PC4 加入 VLAN 200,如图 6-4 所示,则交换机 SW2 如何修改配置?

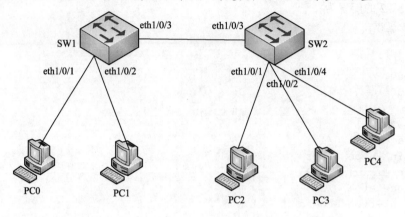

图 6-4 思考题的网络拓扑设计

实验 7　VTP 配置实验（思科模拟器）

背景介绍

VTP（VLAN Trunking Protocol，VLAN 中继协议），又称虚拟局域网干道协议。它是思科私有协议。作用是十几台交换机在企业网中，配置 VLAN 工作量大，可以使用 VTP 协议，把一台交换机配置成 VTP Server，其余交换机配置成 VTP Client，这些 Client 可以自动学习到 Server 上的 VLAN 信息。

实验目的

掌握 VTP 的配置方法，了解 VTP 的工作原理，了解服务器模式、客户端模式和透明模式的特性。

实验步骤

1. 网络拓扑设计

网络拓扑设计如图 7-1 所示。

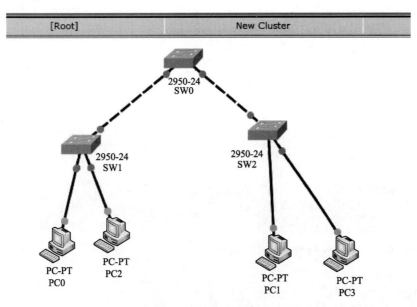

图 7-1　网络拓扑设计

SW0 作为 Server，SW1、SW2 作为 Client。SW0、SW1、SW2 分别配置 VTP 域和 VTP 密码，SW0 划分两个 VLAN，SW1、SW2 不能划分 VLAM，用于转发 VLAN 信息。SW0 配置两个中继端口，SW1、SW2 与 SW0 配置中继。

2. 网络地址规划

网络地址规划如表 7-1 所示。

表 7-1 网络地址规划

设 备	IP 地 址	所属 VLAN
PC0	192.168.1.1/24	VLAN 100
PC1	192.168.2.1/24	VLAN 200
PC2	192.168.1.2/24	VLAN 100
PC3	192.168.2.2/24	VLAN 200

3. 实验配置命令

SW0 的配置为:
```
enable
configure terminal
//配置 vtp
vtp mode server
vtp domain test
vtp password cisco
//配置 vlan
vlan 100
exit
vlan 200
exit
//配置端口
interface f0/1
switchport mode trunk
switchport trunk allowed vlan all
exit
interface f0/2
switchport mode trunk
switchport trunk allowed vlan all
exit
```

SW1 的配置为:
```
enable
configure terminal
//配置 vtp
vtp mode client
vtp domain test
vtp password cisco
//配置端口
interface f0/3
switchport mode trunk
switchport trunk allowed vlan all
exit
interface f0/1
switchport mode access
switchport access allowed vlan 100
exit
interface f0/2
switchport mode access
switchport access allowed vlan 200
exit
```

SW2 的配置为：
```
enable
configure terminal
//配置 vtp
vtp mode client
vtp domain test
vtp password cisco
//配置端口
interface f0/3
switchport mode trunk
switchport trunk allowed vlan all
exit
interface f0/1
switchport mode access
switchport access allowed vlan 100
exit
interface f0/2
switchport mode access
switchport access allowed vlan 200
exit
```

实验验证

同在一个 VLAN 的 PC 可以 ping 通，不同 VLAN 里的 PC 不能 ping 通，如图 7-2 所示。

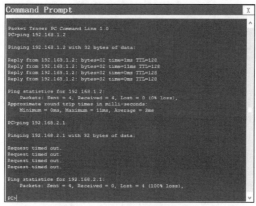

图 7-2　ping 命令测试结果

Client 可以学习到 Server 的 VLAN 信息，如图 7-3 和图 7-4 所示。

图 7-3　学习信息 1

```
%LINEPROTO-5-UPDOWN: Line protocol on Interface FastEthernet0/3, changed state to
up

Switch>
Switch>en
Switch#show vtp st
Switch#show vtp status
VTP Version                    : 2
Configuration Revision         : 2
Maximum VLANs supported locally : 255
Number of existing VLANs       : 7
VTP Operating Mode             : Client
VTP Domain Name                : test
VTP Pruning Mode               : Disabled
VTP V2 Mode                    : Disabled
VTP Traps Generation           : Disabled
MD5 digest                     : 0xA9 0xC4 0x44 0xEE 0xD4 0x27 0xA2 0x70
Configuration last modified by 0.0.0.0 at 3-1-93 00:10:18
Switch#
```

图 7-4　学习信息 2

思考题

请在网络拓扑图中再配置一个透明模式的交换机。

实验 8 VTP 配置实验（上机）

背景介绍

VTP（VLAN Trunking Protocol，VLAN 中继协议），又称虚拟局域网干道协议。它是思科私有协议。作用是十几台交换机在企业网中，配置 VLAN 工作量大，可以使用 VTP 协议，把一台交换机配置成 VTP Server，其余交换机配置成 VTP Client，这些 Client 可以自动学习到 Server 上的 VLAN 信息。

实验目的

掌握 VTP 的配置方法，了解 VTP 的工作原理，了解服务器模式、客户端模式和透明模式的特性。

实验步骤

1. 网络拓扑设计

网络拓扑设计如图 8-1 所示。

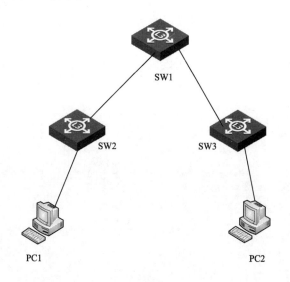

图 8-1 网络拓扑设计

本实验用到了三台 H3C s3100v2-26TP-EI 交换机。SW1 需要与 SW2、SW3 配置中继，并配置 gvrp，SW1、SW2 划分两个 VLAN，与 SW1 配置中继接口并配置 gvrp，最后只需检验 SW1

有没有学习到 SW1 和 SW2 的 VLAN 信息。

2. 网络地址规划

网络地址规划如表 8-1 所示。

表 8-1 网络地址规划

设 备	IP 地 址
PC1	192.168.1.2/24
PC2	192.168.2.2/24

3. 网络配置

SW1 的配置为：
```
sys
//配置端口并全局配置 gvrp
gvrp
interface eth1/0/1
port link-type trunk
port trunk permit vlan all
gvrp
quit
interface eth1/0/2
port link-type trunk
port trunk permit vlan all
gvrp
quit
```
SW2 的配置为：
```
sys
vlan 100
quit
//配置端口并全局配置 gvrp
gvrp
interface eth1/0/1
port link-type trunk
port trunk permit vlan all
gvrp
quit
interface eth1/0/2
port link-type access
port access vlan 100
quit
```
SW3 的配置为：
```
sys
vlan 200
quit
//配置端口并全局配置 gvrp
gvrp
interface eth1/0/1
port link-type trunk
```

```
port trunk permit vlan all
gvrp
quit
interface eth1/0/2
port link-type access
port access vlan 200
quit
```

实验验证

在 SW1 中用命令：dis vlan all，实验结果如图 8-2 所示。

图 8-2　实验结果

思考题

请在网络拓扑图中再加两台 PC，每个 VLAN 里各加一台，然后将这四台 PC 互 ping 一下，看看实验结果。

实验 9

➡ 静态路由（思科模拟器）

背景介绍

静态路由是指由用户或网络管理员手工配置的路由信息。当网络的拓扑结构或链路的状态发生变化时，网络管理员需要手工去修改路由表中相关的静态路由信息。静态路由信息在默认情况下是私有的，不会传递给其他的路由器。当然，网络管理员也可以对路由器进行设置，使之成为共享的。静态路由一般适用于比较简单的网络环境，在这样的环境中，网络管理员易于清楚地了解网络的拓扑结构，便于设置正确的路由信息。

实验目的

掌握路由器的基本配置命令和配置静态路由协议的方法。

实验步骤

1. 网络拓扑设计

实验需要的设备为两台 2621 路由器，两台 PC。网络拓扑设计如图 9-1 所示。

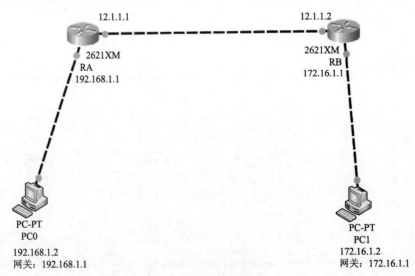

图 9-1　网络拓扑设计

2. 网络和服务配置

各个设备及端口的 IP 地址分配如表 9-1 所示。

表 9-1 IP 地址分配

设 备	接 口	IP 地 址	掩 码	网 关
PC0	Fa0	192.168.1.2	255.255.255.0	192.168.1.1
RA	Fa0/0	192.168.1.1	255.255.255.0	无
RA	Fa0/1	12.1.1.1	255.255.255.0	无
RB	Fa0/1	12.1.1.2	255.255.255.0	无
RB	Fa0/0	172.16.1.1	255.255.255.0	无
PC1	Fa0	172.16.1.2	255.255.255.0	172.16.1.1

路由器 RA 的配置为：
```
//进入特权查看模式
enable
//进入配置模式
conf terminal
//设置路由器名称
hostname RA
//设置 FastEthern0/0 端口信息
interface fastethern0/0
ip address 192.168.1.1 255.255.255.0
no shutdown
exit
//设置 FastEthern0/1 端口信息
interface fastethern0/1
ip address 12.1.1.1 255.255.255.0
no shutdown
exit
//设置静态路由
ip route 172.16.1.0 255.255.255.0 12.1.1.2
```

路由器 RB 的配置为：
```
//设置 FastEthern0/0 端口信息
interface fastethern0/0
ip address 172.16.1.1 255.255.255.0
no shutdown
exit
//设置 FastEthern0/1 端口信息
interface fastethern0/1
ip address 12.1.1.2 255.255.255.0
no shutdown
exit
//设置静态路由
ip route 192.168.1.0 255.255.255.0 12.1.1.1
```

实验验证

从 PC0 ping RA、RB、PC1 的实验结果如图 9-2 所示。

```
PC>ping 192.168.1.1

Pinging 192.168.1.1 with 32 bytes of data:

Reply from 192.168.1.1: bytes=32 time=0ms TTL=255
Reply from 192.168.1.1: bytes=32 time=0ms TTL=255
Reply from 192.168.1.1: bytes=32 time=0ms TTL=255
Reply from 192.168.1.1: bytes=32 time=0ms TTL=255

Ping statistics for 192.168.1.1:
    Packets: Sent = 4, Received = 4, Lost = 0 (0% loss),
Approximate round trip times in milli-seconds:
    Minimum = 0ms, Maximum = 0ms, Average = 0ms

PC>ping 12.1.1.1

Pinging 12.1.1.1 with 32 bytes of data:

Reply from 12.1.1.1: bytes=32 time=0ms TTL=255
Reply from 12.1.1.1: bytes=32 time=0ms TTL=255
Reply from 12.1.1.1: bytes=32 time=0ms TTL=255
Reply from 12.1.1.1: bytes=32 time=0ms TTL=255

Ping statistics for 12.1.1.1:
    Packets: Sent = 4, Received = 4, Lost = 0 (0% loss),
Approximate round trip times in milli-seconds:
    Minimum = 0ms, Maximum = 0ms, Average = 0ms

PC>ping 172.16.1.1

Pinging 172.16.1.1 with 32 bytes of data:

Reply from 172.16.1.1: bytes=32 time=0ms TTL=254
Reply from 172.16.1.1: bytes=32 time=0ms TTL=254
Reply from 172.16.1.1: bytes=32 time=0ms TTL=254
Reply from 172.16.1.1: bytes=32 time=0ms TTL=254

Ping statistics for 172.16.1.1:
    Packets: Sent = 4, Received = 4, Lost = 0 (0% loss),
Approximate round trip times in milli-seconds:
    Minimum = 0ms, Maximum = 0ms, Average = 0ms

PC>ping 172.16.1.2

Pinging 172.16.1.2 with 32 bytes of data:

Reply from 172.16.1.2: bytes=32 time=0ms TTL=126
Reply from 172.16.1.2: bytes=32 time=0ms TTL=126
Reply from 172.16.1.2: bytes=32 time=0ms TTL=126
Reply from 172.16.1.2: bytes=32 time=0ms TTL=126

Ping statistics for 172.16.1.2:
    Packets: Sent = 4, Received = 4, Lost = 0 (0% loss),
Approximate round trip times in milli-seconds:
    Minimum = 0ms, Maximum = 0ms, Average = 0ms

PC>
```

图 9-2　PC0 实验结果

从 PC1 ping RA、RB、PC0 的实验结果如图 9-3 所示。

图 9-3　PC1 实验结果

如果图 9-1 所示的拓扑图中路由器 RB 中增加一个相连的 PC2，其 IP 地址为 10.0.0.240/24，网关为 10.0.0.1/24，如果 PC1、PC2 和 PC0 需要相互访问，则如何修改 RA 和 RB 路由器的配置。

实验⑩ 静态路由（上机）

背景介绍

静态路由是指由用户或网络管理员手工配置的路由信息。当网络的拓扑结构或链路的状态发生变化时，网络管理员需要手工去修改路由表中相关的静态路由信息。静态路由信息在默认情况下是私有的，不会传递给其他的路由器。当然，网络管理员也可以对路由器进行设置，使之成为共享的。静态路由一般适用于比较简单的网络环境，在这样的环境中，网络管理员易于清楚地了解网络的拓扑结构，便于设置正确的路由信息。

实验目的

掌握 H3C 设备中配置静态路由协议的方法。

实验步骤

1. 网络拓扑设计

实验需要两台 H3C MSR36-20 路由器。网络拓扑设计如图 10-1 所示。

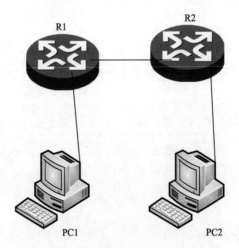

图 10-1 网络拓扑设计

2. 网络配置

各个设备及端口的 IP 地址分配如表 10-1 所示。

表 10-1 IP 地址分配

设备	接口	IP 地址	掩码	网关
PC1	Fa0	192.168.1.2	255.255.255.0	192.168.1.1
R1	Interface g0/0	192.168.1.1	255.255.255.0	无
R1	Interface g0/1	192.168.3.1	255.255.255.0	无
R2	Interface g0/0	192.168.2.1	255.255.255.0	无
R2	Interface g0/1	192.168.3.2	255.255.255.0	无
PC2	Fa0	192.168.2.2	255.255.255.0	192.168.2.1

R1 的配置为：
```
//进入配置模式
sys
//配置端口
interface g0/0
ip address 192.168.1.1 24
quit
interface g0/1
ip address 192.168.3.1 24
quit
//配置静态路由
ip route-static 192.168.2.0 255.255.255.0 192.168.3.2
```
R2 的配置为：
```
//进入配置模式
sys
//配置端口
interface g0/0
ip address 192.168.2.1 24
quit
interface g0/1
ip address 192.168.3.2 24
quit
//配置静态路由
ip route-static 0.0.0.0 0.0.0.0 192.168.3.1
```

实验验证

从 PC1 ping PC2 的实验结果如图 10-2 所示。

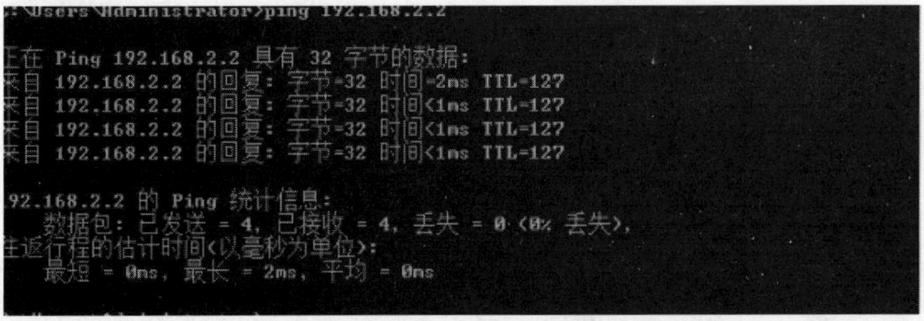

图 10-2 PC1 ping PC2 实验结果

从 PC2 ping PC1 的实验结果如图 10-3 所示。

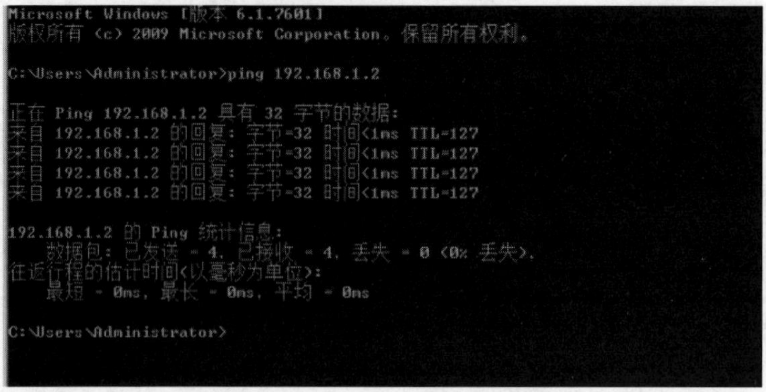

图 10-3　PC2 ping PC1 实验结果

思考题

简述静态路由的实验原理。

实验 11 单臂路由器配置（思科模拟器）

背景介绍

单臂路由（router-on-a-stick）是指在路由器的一个接口上通过配置子接口（或"逻辑接口"，并不存在真正物理接口）的方式，实现原来相互隔离的不同 VLAN（虚拟局域网）之间的互连互通。VLAN 能有效分割局域网，实现各网络区域之间的访问控制。但现实中，往往需要配置某些 VLAN 之间的互连互通。例如，你的公司划分为领导层、销售部、财务部、人力部、科技部、审计部，并为不同部门配置了不同的 VLAN，部门之间不能相互访问，有效保证了各部门的信息安全。但经常出现领导层需要跨越 VLAN 访问其他各个部门，这个功能就由单臂路由来实现。

实验目的

掌握单臂路由器的配置命令，掌握单臂路由器的工作原理。

实验步骤

1. 网络拓扑设计

网络拓扑设计如图 11-1 所示。

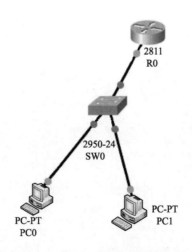

图 11-1 网络拓扑设计

PC0 与 PC1 配置 IP 地址与网关，交换机需要配置中继接口，划分两个不同 VLAN，路由器只需要配置子接口。检验时，两台终端可以互 ping。

2. 网络地址

具体地址规划如表 11-1 所示。

表 11-1 网络地址规划

设 备	接 口	IP 地 址	网 关
PC0	无	192.168.1.2/24	192.168.1.1
PC1	无	192.168.2.2/24	192.168.2.1
R0	f0/0.100	192.168.1.1/24	无
R0	f0/0.200	192.168.2.1/24	无

3. 具体配置

SW1 的配置为：
```
//进入特权模式
en
//进入配置模式
configure terminal
//划分vlan
vlan 100
quit
vlan 200
quit
//配置中继和接入端口
interface f0/1
switchport mode trunk
switchport trunk permit vlan all
exit
interface f0/2
switchport mode access
switchport access vlan 100
exit
interface f0/3
switchport mode access
switchport access vlan 200
exit
```
R0 的配置为：
```
//进入特权模式
en
//进入配置模式
configure terminal
//配置子接口
interface f0/0
no shut
exit
interface f0/0.100
ip address 192.168.1.1 255.255.255.0
encapsulation dot1Q 100
exit
interface f0/0.200
ip address 192.168.2.1 255.255.255.0
```

```
encapsulation dot1Q 200
exit
```

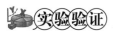

实验验证

PC0 ping PC1 结果如图 11-2 所示。

```
PC>ping 192.168.2.2

Pinging 192.168.2.2 with 32 bytes of data:

Reply from 192.168.2.2: bytes=32 time=0ms TTL=127
Reply from 192.168.2.2: bytes=32 time=3ms TTL=127
Reply from 192.168.2.2: bytes=32 time=0ms TTL=127
Reply from 192.168.2.2: bytes=32 time=0ms TTL=127

Ping statistics for 192.168.2.2:
    Packets: Sent = 4, Received = 4, Lost = 0 (0% loss),
Approximate round trip times in milli-seconds:
    Minimum = 0ms, Maximum = 3ms, Average = 0ms

PC>
```

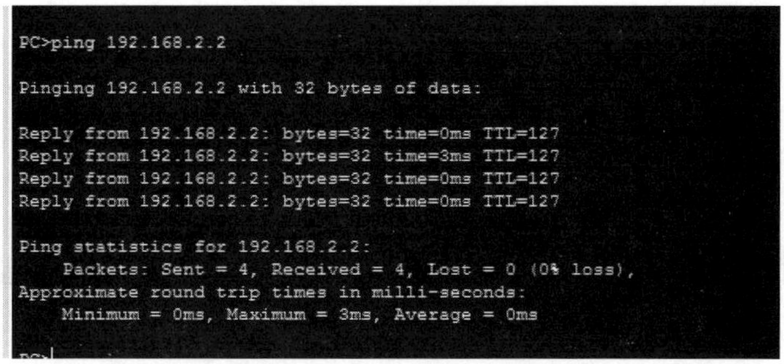

图 11-2　PC0 ping PC1 结果

PC1 ping PC0 结果如图 11-3 所示。

```
Packet Tracer PC Command Line 1.0
PC>ping 192.168.1.2

Pinging 192.168.1.2 with 32 bytes of data:

Reply from 192.168.1.2: bytes=32 time=0ms TTL=127
Reply from 192.168.1.2: bytes=32 time=13ms TTL=127
Reply from 192.168.1.2: bytes=32 time=2ms TTL=127
Reply from 192.168.1.2: bytes=32 time=0ms TTL=127

Ping statistics for 192.168.1.2:
    Packets: Sent = 4, Received = 4, Lost = 0 (0% loss),
Approximate round trip times in milli-seconds:
    Minimum = 0ms, Maximum = 13ms, Average = 3ms

PC>
```

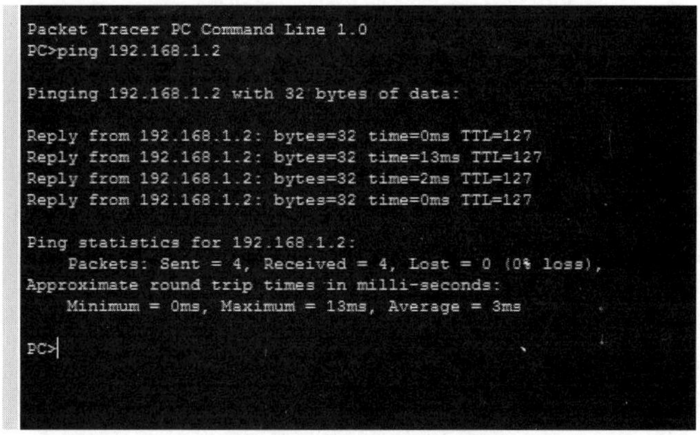

图 11-3　PC1 ping PC0 结果

思考题

请将路由器换成三层交换机来进行实验。

实验 12

➡ 单臂路由器（上机 路由器）

📞 背景介绍

单臂路由（router-on-a-stick）是指在路由器的一个接口上通过配置子接口（或"逻辑接口"，并不存在真正物理接口）的方式，实现原来相互隔离的不同 VLAN（虚拟局域网）之间的互连互通。

⏳ 实验目的

掌握单臂路由器的配置方法，了解子接口的工作原理。

👆 实验步骤

1. 网络拓扑设计

本实验用到了一台 H3C MSR36-20 路由器和一台 H3C s3100v2-26TP-EI 交换机。SW1 划分两个 VLAN，PC1 对应 VLAN100，PC2 对应 VLAN200，并配置中继接口，R1 配置子接口。检验时，PC1 能 ping 通 PC2 即可。网络拓扑设计如图 12-1 所示。

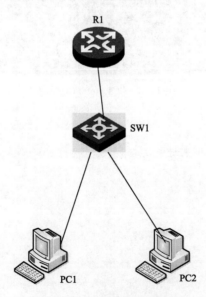

图 12-1 网络拓扑设计

2. 网络地址

网络地址规划如表 12-1 所示。

表 12-1 网络地址规划

设 备	接 口	IP 地 址	网 关
R1	G0/0.100	192.168.1.1/24	无
R2	G0/0.200	192.168.2.1/24	无
PC1	无	192.168.1.2/24	192.168.1.1
PC2	无	192.168.2.2/24	192.168.2.1

3. 具体网络配置

PC1、PC2 配置端口和网关即可。
SW1 的配置为：
```
//进入配置模式
sys
//划分vlan
vlan 100
quit
vlan 200
quit
//配置端口
interface eth1/0/1
port link-type trunk
port trunk permit vlan all
quit
interface eth1/0/2
port link-type access
port access vlan 100
quit
quit
interface eth1/0/3
port link-type access
port access vlan 200
quit
```
R1 的配置为：
```
//进入配置端口
sys
//配置子接口
interface g0/0.100
ip address 192.168.1.1 24
vlan-type dot1q vid 100
quit
interface g0/0.200
ip address 192.168.2.1 24
vlan-type dot1q vid 200
quit
```

实验验证

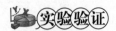

网络验证

PC1 ping PC2 的结果如图 12-2 所示。

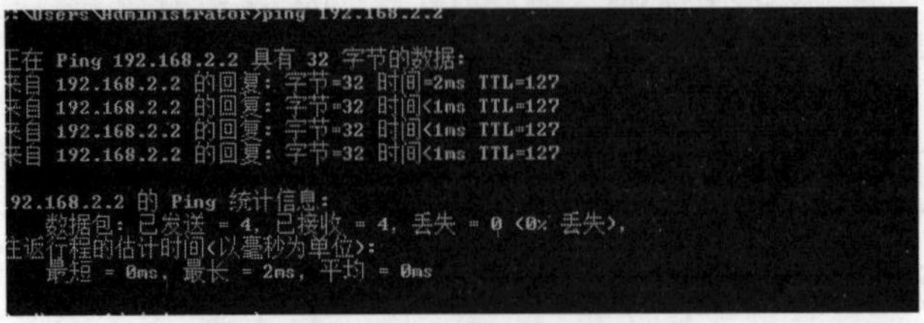

图 12-2 PC1 ping PC2

PC2 ping PC1 的结果如图 12-3 所示。

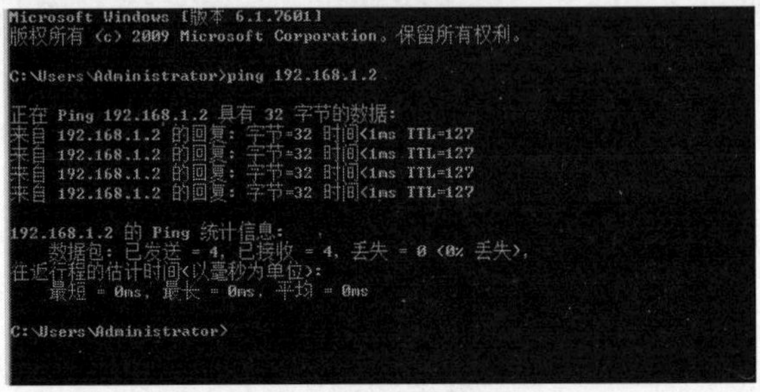

图 12-3 PC2 ping PC1

思考题

请增加一个汇聚交换机来进行这个单臂路由器的实验。

实验 13

单臂路由器（上机 三层交换机）

背景介绍

单臂路由（router-on-a-stick）是指在路由器的一个接口上通过配置子接口（或"逻辑接口"，并不存在真正物理接口）的方式，实现原来相互隔离的不同 VLAN（虚拟局域网）之间的互连互通。

实验目的

掌握单臂路由器的配置方法，了解子接口的工作原理。

实验步骤

1. 网络拓扑设计

本实验用到一台 H3C s3600v2-28TP-EI 三层交换机和一台 H3C s3100v2-26TP-EI 二层交换机。SW1 划分两个 VLAN，PC1 对应 VLAN100，PC2 对应 VLAN200，并配置中继接口，R1 配置子接口。检验时，PC1 能 ping 通 PC2 即可。网络拓扑设计如图 13-1 所示。

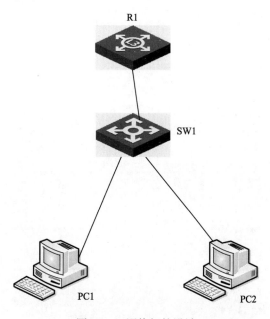

图 13-1 网络拓扑设计

2. 网络地址

具体地址规划如表 13-1 所示。

表 13-1 网络地址规划

设 备	接 口	IP 地 址	网 关
三层交换机	VLAN 100	192.168.1.1/24	无
三层交换机	VLAN 200	192.168.2.1/24	无
三层交换机	Eth1/0/1	无	无
PC1	无	192.168.1.2/24	192.168.1.1
PC2	无	192.168.2.2/24	192.168.2.1

3. 具体网络配置

PC1、PC2 配置端口和网关即可。
SW1 的配置为:
```
//进入配置模式
sys
//划分vlan
vlan 100
quit
vlan 200
quit
//配置端口和接入端口
interface eth1/0/1
port link-type trunk
port trunk permit vlan all
quit
interface eth1/0/2
port link-type access
port access vlan 100
quit
quit
interface eth1/0/3
port link-type access
port access vlan 200
quit
```
三层交换机配置为:
```
//进入配置端口
sys
//划分vlan
vlan 100
quit
vlan 200
quit
interface eth1/0/1
port link-type trunk
port trunk permit vlan all
//配置子接口
interface vlan-interface 100
```

```
ip address 192.168.1.1 24
quit
interface vlan-interface 200
ip address 192.168.2.1 24
quit
```

网络验证

PC1 ping PC2 结果如图 13-2 所示。

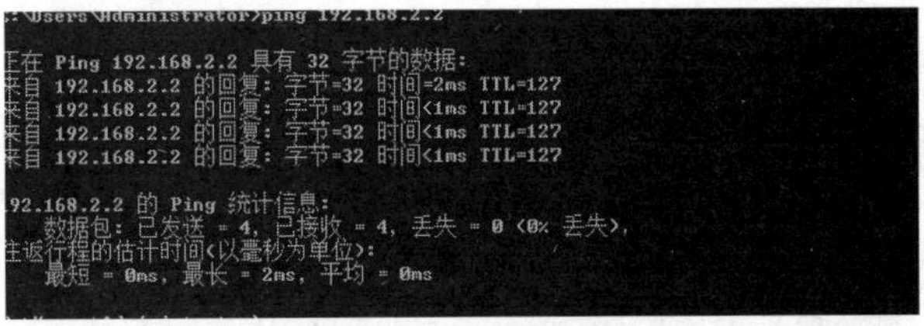

图 13-2　PC1 ping PC2

PC2 ping PC1 结果如图 13-3 所示。

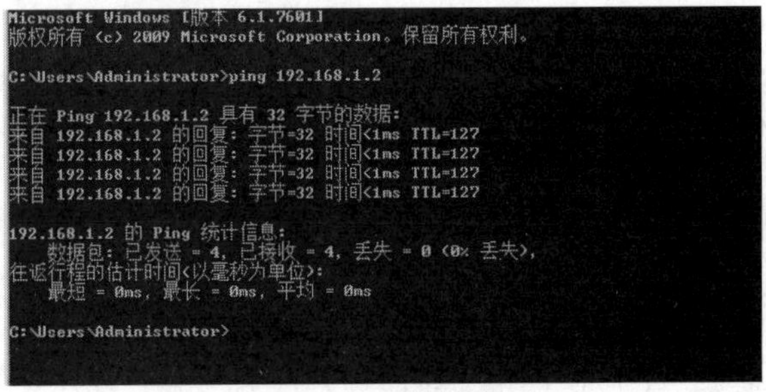

图 13-3　PC2 ping PC1

请增加一个汇聚交换机来做这个单臂路由器的实验。

实验 14

→ RIP 路由（思科模拟器）

背景介绍

RIP（Routing Information Protocol，路由信息协议）是应用较早、使用较普遍的内部网关协议（Interior Gateway Protocol，IGP），适用于小型同类网络的一个自治系统（AS）内的路由信息的传递。RIP 协议是基于距离矢量算法（Distance Vector Algorithms，DVA）的。它使用"跳数"，即 metric 来衡量到达目标地址的路由距离。文档见 RFC1058、RFC1723。它是一个用于路由器和主机间交换路由信息的距离向量协议，目前最新的版本为 v4，也就是 RIPv4。

实验目的

掌握 RIP 路由器的基本设置方法，了解 RIP 路由器基本原理。本实验通过思科模拟器实现。

实验步骤

1. 网络拓扑设计

实验所需要的设备为三台 1841 路由器，两台 PC。网络拓扑设计如图 14-1 所示。

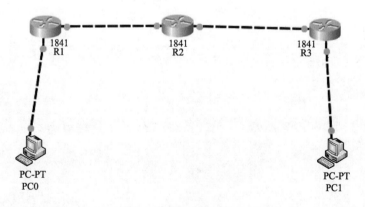

图 14-1 网络拓扑设计

2. 网络和服务器配置

各个设备及端口的 IP 地址分配如表 14-1 所示。

表 14-1　IP 地址分配

设　备	接　口	IP 地　址	掩　码	网　关
R1	FastEthernet0/0	192.168.2.1	255.255.255.0	无
R1	FastEthernet0/1	192.168.1.2	255.255.255.0	无
R2	FastEthernet0/0	192.168.2.2	255.255.255.0	无
R2	FastEthernet0/1	192.168.3.1	255.255.255.0	无
R3	FastEthernet1/0	192.168.3.2	255.255.255.0	无
R3	FastEthernet1/1	192.168.4.2	255.255.255.0	无
PC0	无	192.168.1.1	255.255.255.0	192.168.1.2
PC1	无	192.168.4.1	255.255.255.0	192.168.4.2

3. 具体网络配置

R1 的配置为：
```
//进入特权模式
en
//进入配置模式
configure terminal
//配置端口
interface f0/0
ip address 192.168.2.1 255.255.255.0
no shut
exit
interface f0/1
ip address 192.168.1.2 255.255.255.0
no shut
exit
//配置 RIP
route rip
network 192.168.1.0
network 192.168.2.0
exit
```
R2 的配置为：
```
//进入特权模式
en
//进入配置模式
configure terminal
//配置端口
interface f0/0
ip address 192.168.2.2 255.255.255.0
no shut
exit
interface f0/1
ip address 192.168.3.1 255.255.255.0
no shut
exit
//配置 RIP
route rip
network 192.168.3.0
```

```
network 192.168.2.0
exit
```
R3 的配置为：
```
//进入特权模式
en
//进入配置模式
configure terminal
//配置端口
interface f0/0
ip address 192.168.3.2 255.255.255.0
no shut
exit
interface f0/1
ip address 192.168.4.2 255.255.255.0
no shut
exit
//配置 RIP
route rip
network 192.168.3.0
network 192.168.4.0
exit
```
PC0 终端只需要设置网络信息，如图 14-2 所示。

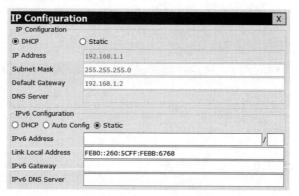

图 14-2　PC0 网络配置

PC1 终端只需要设置网络信息，如图 14-3 所示。

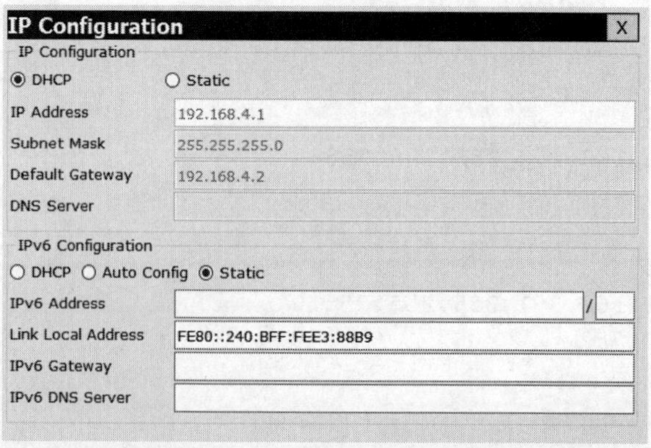

图 14-3　PC1 网络配置

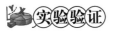

实验验证

从 PC0 ping PC1 的 IP 地址正常，如图 14-4 所示。

```
PC>ping 192.168.4.1

Pinging 192.168.4.1 with 32 bytes of data:

Reply from 192.168.4.1: bytes=32 time=11ms TTL=125
Reply from 192.168.4.1: bytes=32 time=0ms TTL=125
Reply from 192.168.4.1: bytes=32 time=0ms TTL=125
Reply from 192.168.4.1: bytes=32 time=0ms TTL=125

Ping statistics for 192.168.4.1:
    Packets: Sent = 4, Received = 4, Lost = 0 (0% loss),
Approximate round trip times in milli-seconds:
    Minimum = 0ms, Maximum = 11ms, Average = 2ms
```

图 14-4 网络验证结果

思考题

运行 RIP 的路由器在从通告中收到新路由的信息后，首先会做什么？

实验 15 RIP 配置实验（上机）

背景介绍

RIP 协议是一种内部网关协议（IGP），是一种动态路由选择协议，用于自治系统（AS）内的路由信息的传递。

RIP 协议基于距离矢量算法（DistanceVectorAlgorithms），使用"跳数"（即 metric）来衡量到达目标地址的路由距离。这种协议的路由器只关心自己周围的世界，只与自己相邻的路由器交换信息，范围限制在 15 跳（15 度）之内，再远，它就不关心了。RIP 应用于 OSI 网络七层模型的网络层。各厂家定义的管理距离（AD，即优先级）如下：华为定义的优先级是 100，思科定义的优先级是 120。

实验目的

掌握 RIP 的配置，了解 RIP 的工作原理。

实验步骤

1. 网络拓扑设计

本实验的两台设备都是 H3C MSR36-20 路由器。PC1、PC2 只需配置 IP 地址与网关，R1 与 R2 要配置端口与 RIP。检验时，PC1 能 ping 通 PC2。网络拓扑设计如图 15-1 所示。

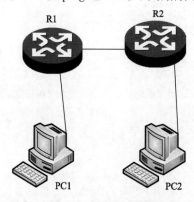

图 15-1 网络拓扑设计

2. 具体网络地址规划

网络地址规划如表 15-1 所示。

表 15-1 网络地址规划

设 备	接 口	IP 地 址	网 关
PC1	无	192.168.1.2/24	192.168.1.1
PC2	无	192.168.2.2/24	192.168.2.1
R1	G0/0	192.168.1.1/24	无
R1	G0/1	192.168.3.1/24	无
R2	G0/0	192.168.2.1/24	无
R2	G0/1	192.168.3.2/24	无

3. 具体网络配置

R1 的配置为：
```
//进入配置模式
sys
//配置端口
interface g0/0
ip address 192.168.1.1 24
quit
interface g0/1
ip address 192.168.3.1 24
quit
//配置 rip
rip
network 192.168.1.0
network 192.168.3.0
quit
```
R2 的配置为：
```
//进入配置模式
sys
//配置端口
interface g0/0
ip address 192.168.2.1 24
quit
interface g0/1
ip address 192.168.3.2 24
quit
//配置 rip
rip
network 192.168.2.0
network 192.168.3.0
quit
```

 实验验证

网络验证

PC1 ping PC2,如图 15-2 所示。

图 15-2　PC1 ping PC2 实验结果

PC2 ping PC1,如图 15-3 所示。

图 15-3　PC2 ping PC1 实验结果

思考题

请用静态路由来实现上述配置。

实验 16 EIGRP 路由（思科模拟器）

背景介绍

EIGRP 是 Cisco 的私有路由协议，它综合了距离矢量和链路状态两者的优点。

实验目的

掌握 EIGRP 协议。

实验步骤

1. 网络拓扑设计

实验所需要的设备为五台 2811 路由器，一台交换机，一台服务器，两台 PC。网络拓扑设计如图 16-1 所示。

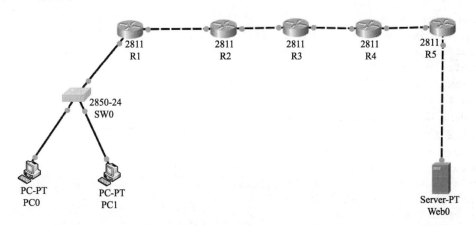

图 16-1　网络拓扑设计

2. 网络和服务器配置

各个设备及端口的 IP 地址分配如表 16-1 所示。

表 16-1　IP 地址分配

设备	接口	IP 地址	掩码	网关	DNS
R1	FastEthernet0/0.100	202.102.216.1	255.255.255.252	无	无
R1	FastEthernet0/1.200	202.102.217.1	255.255.255.252	无	无
R2	FastEthernet0/0	10.0.0.13	255.255.255.252	无	无
R2	FastEthernet0/1	10.0.0.10	255.255.255.252	无	无

设 备	接 口	IP 地址	掩 码	网 关	DNS
R3	FastEthernet0/0	10.0.0.9	255.255.255.252	无	无
R3	FastEthernet0/1	10.0.0.6	255.255.255.252	无	无
R4	FastEthernet0/0	10.0.0.5	255.255.255.252	无	无
R4	FastEthernet0/1	10.0.0.2	255.255.255.252	无	无
R5	FastEthernet0/0	10.0.0.1	255.255.255.252	无	无
R5	FastEthernet0/1	61.132.246.1	255.255.255.252	无	无
PC0	FastEthernet0	202.102.216.2	255.255.255.0	202.102.216.1	无
PC1	FastEthernet0	202.102.217.2	255.255.255.0	202.102.217.1	无

路由器 R1 需要的配置为：

```
//配置端口
interface fastEthernet0/0
 description to-sw1
//配置子接口
interface fastEthernet0/0.100
 encapsulation dot1Q 100
 ip address 202.102.216.1 255.255.255.0
interface fastEthernet0/0.200
 encapsulation dot1Q 200
 ip address 202.102.217.1 255.255.255.0
interface fastEthernet0/1
 description to-r2-fa0/0
 ip address 10.0.0.14 255.255.255.252
//配置eigrp
router eigrp 100
 redistribute connected
 network 10.0.0.0 0.0.0.255
```

路由器 R2 需要的配置为：

```
//配置端口
interface fastEthernet0/0
description to-R1-fa0/1
ip address 10.0.0.13 255.255.255.252
interface fastEthernet0/1
description to-R3-fa0/0
ip address 10.0.0.10 255.255.255.252
interface vlan1
shutdown
//配置eigrp
router eigrp 100
network 10.0.0.0 0.0.0.255
```

路由器 R3 需要的配置为：

```
//配置端口
interface fastEthernet0/0
description to-R2-fa0/1
ip address 10.0.0.9 255.255.255.252
interface fastEthernet0/1
description to-R4-fa0/0
```

```
ip address 10.0.0.6 255.255.255.252
interface vlan1
Shutdown
//配置eigrp
router eigrp 100
network 10.0.0.0 0.0.0.255
```
路由器R4需要的配置为：
```
//配置端口
interface fastEthernet0/0
description to-R3-fa0/1
ip address 10.0.0.5 255.255.255.252
interface fastEthernet0/1
description to-R5-fa0/0
ip address 10.0.0.2 255.255.255.252
interface vlan1
Shutdown
//配置eigrp
router eigrp 100
network 10.0.0.0 0.0.0.255
```
路由器R5需要的配置为：
```
//配置端口
interface fastEthernet0/0
description to-R4-fa0/1
ip address 10.0.0.1 255.255.255.252
interface fastEthernet0/1
description to-serv
ip address 61.132.246.1 255.255.255.252
interface vlan1
shutdown
//配置eigrp
router eigrp 100
redistribute connected
network 10.0.0.0 0.0.0.255
```
交换机SW0需要的配置为：
```
//配置中继和接入端口
interface fastEthernet0/1
switchport mode trunk
interface fastEthernet0/2
switchport access vlan 100
interface fastEthernet0/3
switchport access vlan 200
interface vlan1
shutdown
```

实验验证

网络验证

从PC ping服务器IP地址正常，如图16-2和图16-3所示。

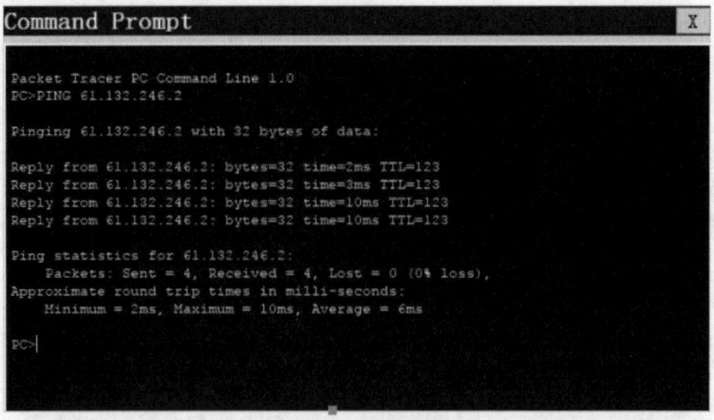

图 16-2　网络验证 1

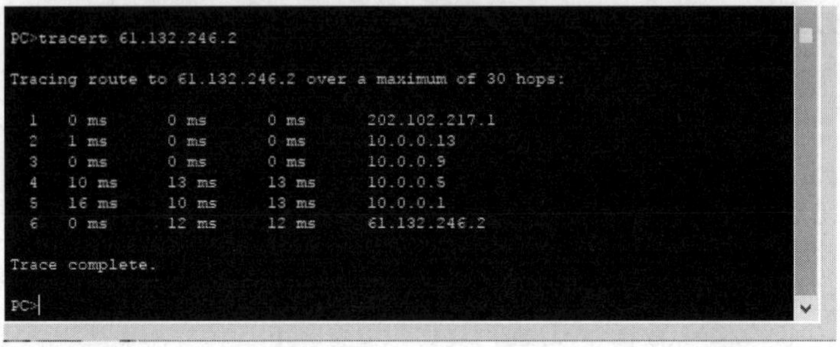

图 16-3　网络验证 2

思考题

EIGRP 路由协议包括哪些数据包？又有哪些特征？

实验 17 OSPF 路由配置实验（思科模拟器）

背景介绍

OSPF（Open Shortest Path First，开放式最短路径优先）是一个内部网关协议（Interior Gateway Protocol，IGP），用于在单一自治系统（Autonomous System，AS）内决策路由，是对链路状态路由协议的一种实现，隶属内部网关协议（IGP），故运作于自治系统内部。著名的迪克斯加算法（Dijkstra）被用来计算最短路径树。OSPF 分为 OSPFv2 和 OSPFv3 两个版本，其中 OSPFv2 用在 IPv4 网络，OSPFv3 用在 IPv6 网络。OSPFv2 是由 RFC 2328 定义的，OSPFv3 是由 RFC 5340 定义的。与 RIP 相比，OSPF 是链路状态协议，而 RIP 是距离矢量协议。

实验目的

本实验通过思科模拟器实现。
（1）学习配置 OSPF 路由。
（2）查看路由器 OSPF 状态。
（3）验证 OSPF 多路径发送数据。

实验步骤

1. 网络拓扑设计

实验所需要的设备为两台 1841 路由器，两根互联线。网络拓扑设计如图 17-1 所示。

图 17-1 网络拓扑设计

2. 网络配置

各个设备及端口的 IP 地址分配如表 17-1 所示。

表 17-1 IP 地址分配

设备	接口	IP 地址	掩码	网关	DNS
R1	FastEthernet0/0	192.168.0.1	255.255.255.252	无	无
R1	FastEthernet0/1	192.168.0.5	255.255.255.252	无	无
R2	FastEthernet0/0	192.168.0.2	255.255.255.252	无	无
R2	FastEthernet0/1	192.168.0.6	255.255.255.252	无	无

R1 和 R2 通过 OSPF 路由协议通信。
路由器 R1 需要的配置为：
```
//进入特权查看模式
enable
//进入配置模式
config ter
hostname R1
//配置端口
interface Loopback0
ip address 1.1.1.1 255.255.255.255
exit
interface FastEthernet0/0
description to-r2-fa0/0
ip address 192.168.0.1 255.255.255.252
ip ospf cost 1000
no shutdown
exit
interface FastEthernet0/1
description to-r2-fa0/1
ip address 192.168.0.5 255.255.255.252
no shutdown
exit
//配置 ospf
router ospf 1
log-adjacency-changes
network 192.168.0.0 0.0.0.255 area 0
network 1.1.1.1 0.0.0.0 area 0
```
路由器 R2 需要的配置为：
```
//进入特权查看模式
enable
//进入特权查看模式
config ter
hostname R2
//配置端口
interface Loopback0
ip address 2.2.2.2 255.255.255.255
exit
interface FastEthernet0/0
description to-r1-fa0/0
ip address 192.168.0.2 255.255.255.252
ip ospf cost 1000
no shutdown
exit
interface FastEthernet0/1
description to-r1-fa0/1
ip address 192.168.0.6 255.255.255.252
ip ospf cost 500
no shutdown
exit
//配置 ospf
router ospf 2
log-adjacency-changes
```

```
network 192.168.0.0 0.0.0.255 area 0.0.0.0
network 2.2.2.2 0.0.0.0 area 0.0.0.0
```

实验验证

OSPF 邻居关系正常，为 FULL，如图 17-2 所示。

```
R1#show ip ospf nei

Neighbor ID     Pri   State          Dead Time   Address         Interface
2.2.2.2           1   FULL/DR        00:00:39    192.168.0.2     FastEthernet0/0
2.2.2.2           1   FULL/DR        00:00:39    192.168.0.6     FastEthernet0/1
R1#
```

图 17-2　OSPF 邻居关系

从 R1 路由器 ping 2.2.2.2 正常，从 R2 ping 1.1.1.1 正常，如图 17-3 所示。

```
R2>ping 1.1.1.1
Type escape sequence to abort.
Sending 5, 100-byte ICMP Echos to 1.1.1.1, timeout is 2 seconds:
!!!!!
Success rate is 100 percent (5/5), round-trip min/avg/max = 0/0/1 ms
```

```
R1#ping 2.2.2.2
Type escape sequence to abort.
Sending 5, 100-byte ICMP Echos to 2.2.2.2, timeout is 2 seconds:
!!!!!
Success rate is 100 percent (5/5), round-trip min/avg/max = 0/0/1 ms
```

图 17-3　网络验证结果

验证传输路径走的是代价小的路径，如图 17-4 所示。

```
R1#tracero 2.2.2.2
Type escape sequence to abort.
Tracing the route to 2.2.2.2

  1   192.168.0.6      1 msec     0 msec     0 msec
```

图 17-4　验证传输路径

查看 OSPF 数据库的内容，如图 17-5 所示。

```
R1#show ip ospf data
         OSPF Router with ID (1.1.1.1) (Process ID 1)

            Router Link States (Area 0)

Link ID         ADV Router      Age         Seq#         Checksum Link count
1.1.1.1         1.1.1.1         470         0x80000006   0x00701f 3
2.2.2.2         2.2.2.2         470         0x80000006   0x0086fa 3

            Net Link States (Area 0)
Link ID         ADV Router      Age         Seq#         Checksum
192.168.0.6     2.2.2.2         475         0x80000003   0x00c70f
192.168.0.2     2.2.2.2         470         0x80000004   0x004899
```

图 17-5　OSPF 数据库

如果改为多路由器不仅仅只有两个路由器的环型链路，又该怎么配置 OSPF 呢？

实验 18 OSPF 路由配置实验（上机）

背景介绍

OSPF（Open Shortest Path First，开放式最短路径优先）是一个内部网关协议（Interior Gateway Protocol，IGP），用于在单一自治系统（Autonomous System，AS）内决策路由，是对链路状态路由协议的一种实现，隶属内部网关协议（IGP），故运作于自治系统内部。著名的迪克斯加算法（Dijkstra）被用来计算最短路径树。OSPF 分为 OSPFv2 和 OSPFv3 两个版本，其中 OSPFv2 用在 IPv4 网络，OSPFv3 用在 IPv6 网络。OSPFv2 是由 RFC 2328 定义的，OSPFv3 是由 RFC 5340 定义的。与 RIP 相比，OSPF 是链路状态协议，而 RIP 是距离矢量协议。

实验目的

本实验通过华为实体设备实现。
（1）学习配置 OSPF 路由。
（2）查看路由器 OSPF 状态。
（3）验证 OSPF 多路径发送数据。

实验步骤

1. 网络拓扑设计

实验需要两台 H3C MSR36-20 路由器。网络拓扑设计如图 18-1 所示。

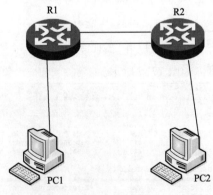

图 18-1 网络拓扑设计

2. 网络配置

各个设备及端口的 IP 地址分配如表 18-1 所示。

实验 18 OSPF 路由配置实验（上机）

表 18-1 IP 地址分配

设 备	接 口	IP 地 址	掩 码	网 关	DNS
R1	FastEthernet0/1	192.168.0.1	255.255.255.252	无	无
R1	FastEthernet0/2	192.168.0.5	255.255.255.252	无	无
R2	FastEthernet0/1	192.168.0.2	255.255.255.252	无	无
R2	FastEthernet0/2	192.168.0.6	255.255.255.252	无	无
R1	FastEthernet0/0	10.0.0.4	255.255.255.0	无	无
R2	FastEthernet0/0	172.16.0.5	255.255.255.0	无	无
PC1	无	10.0.0.1	255.255.255.0	10.0.0.4	无
PC2	无	172.16.0.1	255.255.255.0	172.16.0.5	无

R1 和 R2 通过 OSPF 路由协议通信。
路由器 R1 需要的配置为：
//进入配置模式
sys
//关闭警告信息
undo info enable
sysname R1
//配置 dhcp 地址池
dhcp server ip-pool 10
network 10.0.0.0 mask 255.255.255.0
gateway-list 10.0.0.4
//配置端口
interface GigabitEthernet0/0
port link-mode route
ip address 10.0.0.4 255.255.255.0
undo shut
interface GigabitEthernet0/1
port link-mode route
ip address 192.168.0.1 255.255.255.252
ospf cost 1000
undo shut
interface GigabitEthernet0/2
port link-mode route
ip address 192.168.0.5 255.255.255.252
ospf cost 500
undo shut
quit
//配置 ospf
ospf 1 router-id 10.0.0.4
area 0.0.0.0
network 10.0.0.0 0.0.0.255
network 192.168.0.0 0.0.0.3
network 192.168.0.4 0.0.0.3
路由器 R2 需要的配置为：
//进入配置模式
sys
//关闭警告信息
undo info enable

```
sysname R2
//配置 dhcp 地址池
dhcp server ip-pool 10
network 172.16.0.0 mask 255.255.255.0
gateway-list 172.16.0.5
//配置端口
interface GigabitEthernet0/0
port link-mode route
ip address 172.16.0.5 255.255.255.0
undo shut
interface GigabitEthernet0/1
port link-mode route
ip address 192.168.0.2 255.255.255.252
ospf cost 1000
undo shut
interface GigabitEthernet0/2
port link-mode route
ip address 192.168.0.6 255.255.255.252
ospf cost 500
undo shut
quit
//配置 ospf
ospf 2 router-id 172.16.0.5
area 0.0.0.0
network 172.16.0.0 0.0.0.255
network 192.168.0.0 0.0.0.3
network 192.168.0.4 0.0.0.3
```

实验验证

OSPF 邻居关系正常，为 FULL，如图 18-2 所示。

```
<R1>dis ospf peer

        OSPF Process 1 with Router ID 10.0.0.4
              Neighbor Brief Information

 Area: 0.0.0.0
 Router ID      Address         Pri Dead-Time  State      Interface
 172.16.0.5     192.168.0.2     1   37         Full/BDR   GE0/1
 172.16.0.5     192.168.0.6     1   34         Full/BDR   GE0/2
<R1>
```

图 18-2 OSPF 邻居关系

查看 OSPF 数据库的内容，如图 18-3 所示。

```
[R1]dis ospf lsdb

        OSPF Process 1 with Router ID 10.0.0.4
                 Link State Database

                    Area: 0.0.0.0
 Type      LinkState ID    AdvRouter       Age   Len   Sequence    Metric
 Router    10.0.0.4        10.0.0.4        92    60    8000000C    0
 Router    172.16.0.5      172.16.0.5      316   48    80000006    0
 Network   192.168.0.5     10.0.0.4        302   32    80000002    0
 Network   192.168.0.1     10.0.0.4        312   32    80000002    0
[R1]
```

图 18-3 OSPF 数据库

从 PC1 ping PC2 正常，如图 18-4 所示。

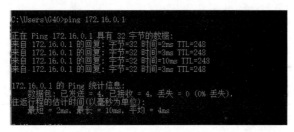

图 18-4　网络验证结果

如果改为多路由器不仅仅只有两个路由器的环型链路，又该怎么配置 OSPF 呢？

实验 19

ISIS 路由配置实验（华为模拟器）

背景介绍

中间系统到中间系统（Intermediate System to Intermediate System，IS-IS）是一种内部网关协议，是电信运营商普遍采用的内部网关协议之一。标准的 IS-IS 协议是由国际标准化组织制定的 ISO/IEC 10589:2002 所规范的。但是标准的 IS-IS 协议是为无连接网络服务（CLNS）设计的，并不直接适合 IP 网络，因此互联网工程任务组制定了可以适用于 IP 网络的集成化的 IS-IS 协议，称为集成 IS-IS，它由 RFC 1195 等 RFC 文档所规范。由于 IP 网络普遍存在，一般所称的 IS-IS 协议，通常是指集成 IS-IS 协议。

实验目的

掌握 IS-IS 的基本配置方法，了解 IS-IS 的基本原理。本实验通过华为模拟器实现。

实验步骤

1. 网络拓扑设计

实验所需要的设备为三台 AR2240 路由器，建议使用 AR2240 路由器，因为其他路由可能不支持本实验，可能会导致实验结果有错误。网络拓扑设计如图 19-1 所示。

图 19-1 网络拓扑设计

2. 网络配置

各个设备及端口的 IP 地址分配如表 19-1 所示。

表 19-1 IP 地址分配

设备	接口	IP 地址	掩码	网关	Loopback 地址
AR1	FastEthernet0/0	192.168.0.1	255.255.255.0	无	1.1.1.1
AR2	FastEthernet0/0	192.168.0.2	255.255.255.0	无	2.2.2.2
AR2	FastEthernet0/1	10.0.0.2	255.255.255.0	无	2.2.2.2
AR3	FastEthernet0/0	10.0.0.1	255.255.255.0	无	3.3.3.3
AR3	FastEthernet0/1	10.1.1.1	255.255.255.0	无	3.3.3.3

实验 19 ISIS 路由配置实验（华为模拟器）

AR1，AR2，AR3 通过 ISIS 路由协议通信。
路由器 AR1 需要的配置为：
//进入配置模式
sys
//关闭警告信息
undo info enable
sysname R1
//配置 isis 和端口
isis 1
is-level level-1
network-entity 10.0001.0010.0100.1001.00
quit
interface GigabitEthernet0/0/0
ip address 192.168.0.1 255.255.255.0
 undo shut
isis enable 1
interface LoopBack0
ip address 1.1.1.1 255.255.255.255
isis enable 1
路由器 AR2 需要的配置为：
//进入配置模式
sys
//关闭警告信息
undo info enable
sysname R2
//配置 isis 和端口
isis 1
is-level level-1-2
network-entity 10.0001.0020.0200.2002.00
quit
interface GigabitEthernet0/0/0
ip address 192.168.0.2 255.255.255.0
 undo shut
isis enable 1
isis circuit-level level-1
interface GigabitEthernet0/0/1
ip address 10.0.0.2 255.255.255.0
 undo shut
isis enable 1
isis circuit-level level-2
interface LoopBack0
ip address 2.2.2.2 255.255.255.255
isis enable 1
路由器 AR3 需要的配置为：
//进入配置模式
sys
//关闭警告信息
undo info enable
sysname R3
//配置 isis 和端口

```
isis 1
is-level level-2
network-entity 10.0002.0030.0300.3003.00
quit
interface GigabitEthernet0/0/0
ip address 10.0.0.1 255.255.255.0
 undo shut
isis enable 1
interface GigabitEthernet0/0/1
ip address 10.1.1.1 255.255.255.0
 undo shut
isis enable 1
interface LoopBack0
ip address 3.3.3.3 255.255.255.255
isis enable 1
```

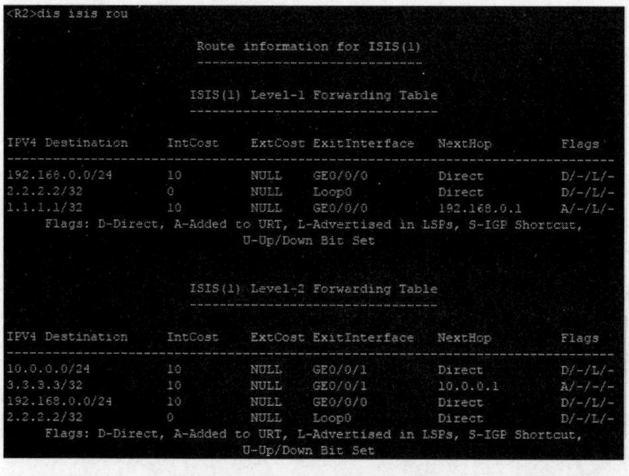

实验验证

展示 ISIS 路由，可使用命令 display isis route ，结果如图 19-2 ~ 图 19-4 所示。

AR1 验证结果：

```
<R1>dis isis rou
                         Route information for ISIS(1)
                         -----------------------------

                        ISIS(1) Level-1 Forwarding Table
                        --------------------------------

IPV4 Destination     IntCost    ExtCost ExitInterface   NextHop         Flags
-------------------------------------------------------------------------------
0.0.0.0/0            10         NULL    GE0/0/0         192.168.0.2     A/-/-/-
192.168.0.0/24       10         NULL    GE0/0/0         Direct          D/-/L/-
2.2.2.2/32           10         NULL    GE0/0/0         192.168.0.2     A/-/-/-
1.1.1.1/32           0          NULL    Loop0           Direct          D/-/L/-
     Flags: D-Direct, A-Added to URT, L-Advertised in LSPs, S-IGP Shortcut,
                               U-Up/Down Bit Set
```

图 19-2 AR1 验证结果

AR2 验证结果：

```
<R2>dis isis rou
                         Route information for ISIS(1)
                         -----------------------------

                        ISIS(1) Level-1 Forwarding Table
                        --------------------------------

IPV4 Destination     IntCost    ExtCost ExitInterface   NextHop         Flags
-------------------------------------------------------------------------------
192.168.0.0/24       10         NULL    GE0/0/0         Direct          D/-/L/-
2.2.2.2/32           0          NULL    Loop0           Direct          D/-/L/-
1.1.1.1/32           10         NULL    GE0/0/0         192.168.0.1     A/-/L/-
     Flags: D-Direct, A-Added to URT, L-Advertised in LSPs, S-IGP Shortcut,
                               U-Up/Down Bit Set

                        ISIS(1) Level-2 Forwarding Table
                        --------------------------------

IPV4 Destination     IntCost    ExtCost ExitInterface   NextHop         Flags
-------------------------------------------------------------------------------
10.0.0.0/24          10         NULL    GE0/0/1         Direct          D/-/L/-
3.3.3.3/32           10         NULL    GE0/0/1         10.0.0.1        A/-/-/-
192.168.0.0/24       10         NULL    GE0/0/0         Direct          D/-/L/-
2.2.2.2/32           0          NULL    Loop0           Direct          D/-/L/-
     Flags: D-Direct, A-Added to URT, L-Advertised in LSPs, S-IGP Shortcut,
                               U-Up/Down Bit Set
```

图 19-3 AR2 验证结果

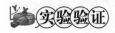

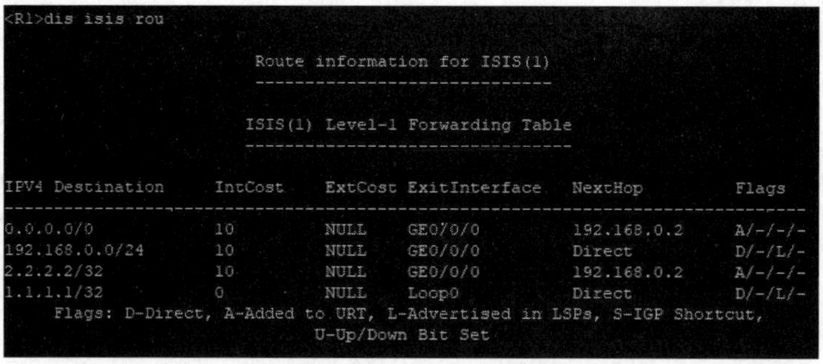

AR3 验证结果：

```
<R3>

  Please check whether system data has been changed, and save data in time

  Configuration console time out, please press any key to log on

<R3>dis isis rou

                         Route information for ISIS(1)
                         -----------------------------

                        ISIS(1) Level-2 Forwarding Table
                        --------------------------------

IPV4 Destination     IntCost    ExtCost ExitInterface    NextHop          Flags
-------------------------------------------------------------------------------
10.0.0.0/24          10         NULL    GE0/0/0          Direct           D/-/L/-
3.3.3.3/32           0          NULL    Loop0            Direct           D/-/L/-
192.168.0.0/24       20         NULL    GE0/0/0          10.0.0.2         A/-/-/-
2.2.2.2/32           10         NULL    GE0/0/0          10.0.0.2         A/-/-/-
1.1.1.1/32           20         NULL    GE0/0/0          10.0.0.2         A/-/-/-
     Flags: D-Direct, A-Added to URT, L-Advertised in LSPs, S-IGP Shortcut,
                               U-Up/Down Bit Set
```

图 19-4 AR3 验证结果

从 AR1 ping 到 AR3，如图 19-5 所示。

```
<R1>ping 3.3.3.3
  PING 3.3.3.3: 56  data bytes, press CTRL_C to break
    Reply from 3.3.3.3: bytes=56 Sequence=1 ttl=254 time=90 ms
    Reply from 3.3.3.3: bytes=56 Sequence=2 ttl=254 time=40 ms
    Reply from 3.3.3.3: bytes=56 Sequence=3 ttl=254 time=10 ms
    Reply from 3.3.3.3: bytes=56 Sequence=4 ttl=254 time=10 ms
    Reply from 3.3.3.3: bytes=56 Sequence=5 ttl=254 time=20 ms

  --- 3.3.3.3 ping statistics ---
    5 packet(s) transmitted
    5 packet(s) received
    0.00% packet loss
    round-trip min/avg/max = 10/34/90 ms
```

图 19-5 AR3 验证结果

思考题

图 19-1 所示是三台路由的简单配置，如果换成多区域并且需设置优先级，又该怎么配置呢？

实验 20 ISIS 路由配置实验（上机）

背景介绍

中间系统到中间系统（Intermediate System to Intermediate System，IS-IS）是一种内部网关协议，是电信运营商普遍采用的内部网关协议之一。标准的 IS-IS 协议是由国际标准化组织制定的 ISO/IEC 10589:2002 所规范的。但是标准的 IS-IS 协议是为无连接网络服务（CLNS）设计的，并不直接适合 IP 网络，因此互联网工程任务组制定了可以适用于 IP 网络的集成化的 IS-IS 协议，称为集成 IS-IS，它由 RFC 1195 等 RFC 文档所规范。由于 IP 网络普遍存在，一般所称的 IS-IS 协议，通常是指集成 IS-IS 协议。

实验目的

掌握 IS-IS 的基本配置方法，了解 IS-IS 的基本原理。本实验通过华为设备实现。

实验步骤

1. 网络拓扑设计

实验所需要的设备为三台 H3C MSR36-20 路由器，建议使用 H3C MSR36-20 路由器，因为其他路由可能不支持本实验，可能会导致实验结果有错误。网络拓扑设计如图 20-1 所示。

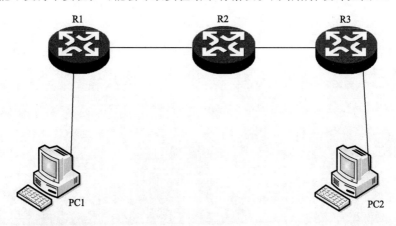

图 20-1　网络拓扑设计

2. 网络配置

各个设备及端口的 IP 地址分配如表 20-1 所示。

表 20-1 IP 地址分配

设 备	接 口	IP 地址	掩 码	网 关	Loopback 地址
R1	FastEthernet0/0	192.168.0.1	255.255.255.0	无	1.1.1.1
R1	FastEthernet0/1	192.168.1.1	255.255.255.0	无	1.1.1.1
R2	FastEthernet0/0	192.168.0.2	255.255.255.0	无	2.2.2.2
R2	FastEthernet0/1	10.0.0.2	255.255.255.0	无	2.2.2.2
R3	FastEthernet0/0	10.0.0.1	255.255.255.0	无	3.3.3.3
R3	FastEthernet0/1	10.0.1.1	255.255.255.0	无	3.3.3.3
PC1	无	192.168.1.2	255.255.255.0	192.168.1.1	无
PC2	无	10.0.1.2	255.255.255.0	10.0.1.1	无

R1，R2，R3 通过 ISIS 路由协议通信。

路由器 R1 需要的配置为：

```
//进入配置模式
sys
//关闭警告信息
undo info enable
sysname R1
//配置 Isis 和端口
isis 1
is-level level-1
network-entity 10.0001.0010.0100.1001.00
quit
interface GigabitEthernet0/0
ip address 192.168.0.1 255.255.255.0
undo shut
isis enable 1
interface GigabitEthernet0/1
ip address 192.168.1.1 255.255.255.0
undo shut
isis enable 1
interface LoopBack0
ip address 1.1.1.1 255.255.255.255
isis enable 1
quit
```

路由器 R2 需要的配置为：

```
//进入配置模式
sys
//关闭警告信息
undo info enable
sysname R2
//配置 Isis 和端口
isis 1
is-level level-1-2
network-entity 10.0001.0020.0200.2002.00
```

```
quit
interface GigabitEthernet0/0
ip address 192.168.0.2 255.255.255.0
undo shut
isis enable 1
isis circuit-level level-1
interface GigabitEthernet0/1
ip address 10.0.0.2 255.255.255.0
undo shut
isis enable 1
isis circuit-level level-2
interface LoopBack0
ip address 2.2.2.2 255.255.255.255
isis enable 1
quit
```

路由器 R3 需要的配置为：

```
//进入配置模式
sys
//关闭警告信息
undo info enable
sysname R3
//配置 Isis 和端口
isis 1
is-level level-2
network-entity 10.0002.0030.0300.3003.00
quit
interface GigabitEthernet0/1
ip address 10.0.0.1 255.255.255.0
undo shut
isis enable 1
interface GigabitEthernet0/0
ip address 10.0.1.1 255.255.255.0
undo shut
isis enable 1
interface LoopBack0
ip address 3.3.3.3 255.255.255.255
isis enable 1
quit
```

实验验证

展示 ISIS 路由，可使用命令 display isis route，结果如图 20-2～图 20-4 所示。

R1 显示：

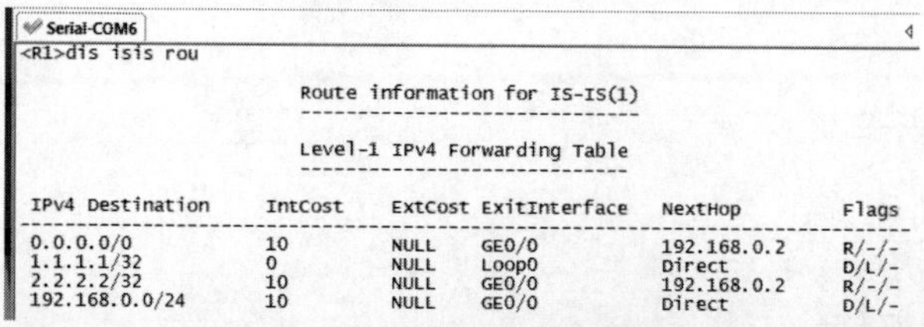

图 20-2　路由 R1 显示

实验 20　ISIS 路由配置实验（上机）　67

R2 显示：

```
<R2>dis isis rou
                  Route information for IS-IS(1)
                  -------------------------------
                  Level-1 IPv4 Forwarding Table
                  -----------------------------
IPv4 Destination    IntCost   ExtCost  ExitInterface    NextHop         Flags
1.1.1.1/32          10        NULL     GE0/0            192.168.0.1     R/L/-
2.2.2.2/32          0         NULL     Loop0            Direct          D/L/-
192.168.0.0/24      10        NULL     GE0/0            Direct          D/L/-
        Flags: D-Direct, R-Added to Rib, L-Advertised in LSPs, U-Up/Down Bit Set
                  Level-2 IPv4 Forwarding Table
                  -----------------------------
IPv4 Destination    IntCost   ExtCost  ExitInterface    NextHop         Flags
10.0.0/24           10        NULL     GE0/1            Direct          D/L/-
2.2.2.2/32          0         NULL                      Direct          D/L/-
3.3.3.3/32          10        NULL     GE0/1            10.0.0.1        R/-/-
192.168.0.0/24      10        NULL                                      D/L/-
```

图 20-3　路由 R2 显示

R3 显示：

```
<R3>dis isis rou
                  Route information for ISIS(1)
                  ------------------------------
                  ISIS(1) IPv4 Level-2 Forwarding Table
                  -------------------------------------
IPv4 Destination    IntCost   ExtCost  ExitInterface    NextHop         Flags
3.3.3.3/32          0         NULL     Loop0            Direct          D/L/-
1.1.1.1/32          20        NULL     Eth0/1           10.0.0.2        R/-/-
10.0.0.0/24         10        NULL     Eth0/1           Direct          D/L/-
192.168.0.0/24      20        NULL     Eth0/1           10.0.0.2        R/-/-
2.2.2.2/32          10        NULL     Eth0/1           10.0.0.2        R/-/-
        Flags: D-Direct, R-Added to RM, L-Advertised in LSPs, U-Up/Down Bit Set
```

图 20-4　路由 R3 显示

从 PC1 ping 到 P3，如图 20-5 所示。

图 20-5　PC1 ping 到 P3

从 PC3 ping 到 P1，如图 20-6 所示。

图 20-6　PC3 ping 到 P1

思考题

图 20-1 所示是三台路由的简单配置，如果换成多区域并且需设置优先级，又该怎么配置呢？

实验 21 BGP 路由配置实验（思科模拟器）

背景介绍

边界网关协议（BGP）是运行于 TCP 上的一种自治系统的路由协议。BGP 是唯一一个用来处理像因特网大小的网络的协议，也是唯一能够妥善处理好不相关路由域间的多路连接的协议。BGP 构建在 EGP 的经验之上。BGP 系统的主要功能是和其他的 BGP 系统交换网络可达信息。网络可达信息包括列出的自治系统（AS）的信息。这些信息有效地构造了 AS 互联的拓扑图并由此清除了路由环路，同时在 AS 级别上可实施策略决策。

实验目的

掌握 BGP 路由协议的基本配置方法，了解 BGP 路由的基本原理。本实验通过思科模拟器实现。

实验步骤

1. 网络拓扑设计

实验所需要的设备为三台 1841 路由器，两台 PC。这里的两台 PC 的 IP 地址采用 DHCP 自动分配地址的方式。网络拓扑设计如图 21-1 所示。

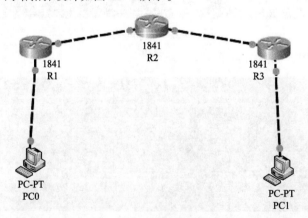

图 21-1　网络拓扑设计

2. 网络配置

各个设备及端口的 IP 地址分配如表 21-1 所示。

表 21-1　IP 地址分配

设备	接口	IP 地址	掩码	网关
R1	FastEthernet0/0	20.1.1.2	255.255.255.0	无
R1	FastEthernet0/1	10.1.1.1	255.255.255.0	无
R2	FastEthernet0/0	20.1.1.1	255.255.255.0	无
R2	FastEthernet0/1	30.1.1.1	255.255.255.0	无
R3	FastEthernet0/0	30.1.1.2	255.255.255.0	无
R3	FastEthernet0/1	40.1.1.1	255.255.255.0	无
PC0	无	10.1.1.2	255.255.255.0	10.1.1.1
PC1	无	40.1.1.2	255.255.255.0	40.1.1.1

R1，R2，R3 通过 BGP 路由协议通信。

路由器 R1 需要的配置为：

```
//进入特权查看模式
enable
//进入配置模式
config ter
hostname R1
//配置 dhcp 地址池
ip dhcp pool 1
network 10.1.1.0 255.255.255.0
default-router 10.1.1.1
exit
//配置端口
interface FastEthernet0/0
ip address 20.1.1.2 255.255.255.0
no shut
exit
interface FastEthernet0/1
ip address 10.1.1.1 255.255.255.0
no shut
exit
//配置 bgp
router bgp 1
bgp log-neighbor-changes
no synchronization
neighbor 20.1.1.1 remote-as 2
network 20.1.1.0mask 255.255.255.0
network 10.1.1.0 mask 255.255.255.0
exit
```

路由器 R2 需要的配置为：

```
//进入特权配置模式
enable
//进入配置模式
config ter
hostname R2
//配置端口
interface FastEthernet0/0
ip address 20.1.1.1 255.255.255.0
no shut
```

```
exit
interface FastEthernet0/1
ip address 30.1.1.1 255.255.255.0
no shut
exit
//配置bgp
router bgp 2
bgp log-neighbor-changes
no synchronization
neighbor 20.1.1.2 remote-as 1
neighbor 20.1.1.2 next-hop-self
neighbor 30.1.1.2 remote-as 3
exit
```

路由器 R3 需要的配置为：

```
//进入特权查看模式
enable
//进入配置模式
config ter
hostname R3
//配置dhcp地址池
ip dhcp pool 1
network 40.1.1.0 255.255.255.0
default-router 40.1.1.1
exit
//配置端口
interface FastEthernet0/0
ip address 30.1.1.2 255.255.255.0
no shut
exit
interface FastEthernet0/1
ip address 40.1.1.1 255.255.255.0
no shut
exit
//配置bgp
router bgp 3
bgp log-neighbor-changes
no synchronization
neighbor 30.1.1.1 remote-as 2
network 40.1.1.0 mask 255.255.255.0
exit
```

PC 的 IP 地址设置如图 21-2 和图 21-3 所示。

PC0 地址设置：

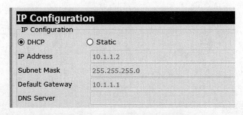

图 21-2　PC0 IP 地址设置

PC1 地址设置：

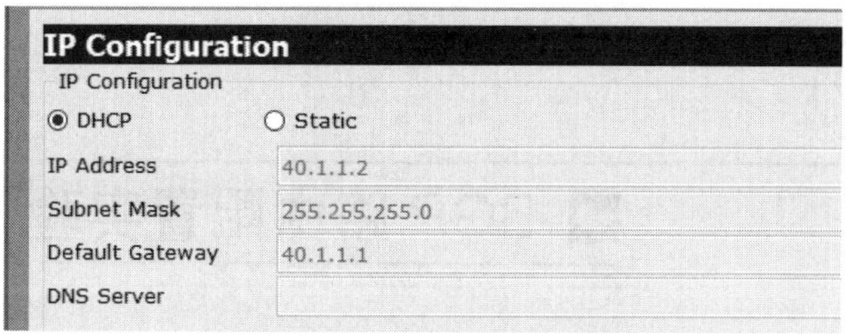

图 21-3　PC1 IP 地址设置

实验验证

从 PC0 ping PC1 地址正常，如图 21-4 所示。

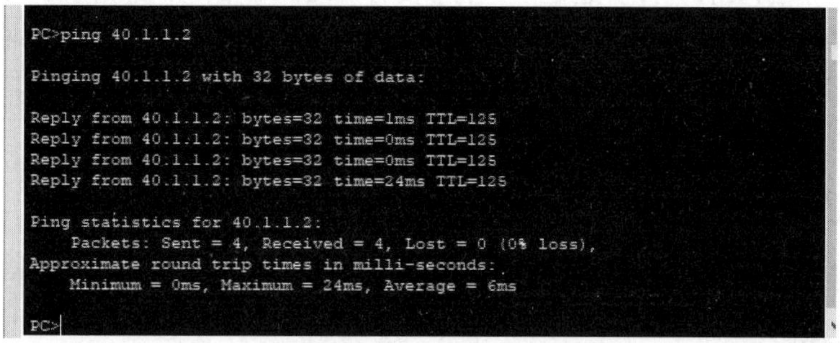

图 21-4　PC0 ping PC1

从 PC1 ping PC0 地址正常，如图 21-5 所示。

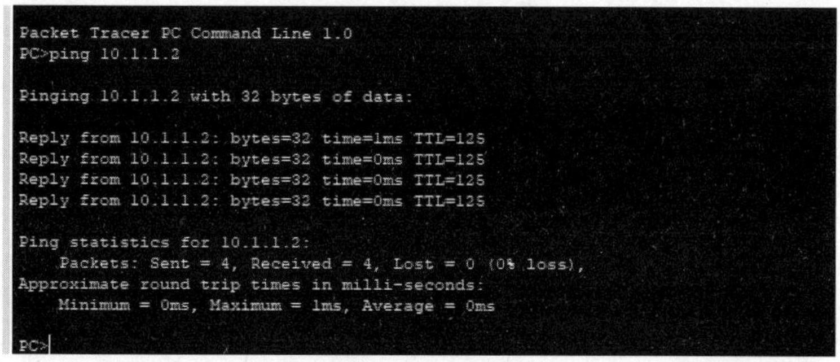

图 21-5　PC1 ping PC0

思考题

如果再添加两个路由器与 R2 路由器分别相连，请问此时 R2 路由器又该怎么对两个新路由器进行配置呢？

实验 22　BGP 路由配置实验（上机）

背景介绍

边界网关协议（BGP）是运行于 TCP 上的一种自治系统的路由协议。BGP 是唯一一个用来处理像因特网大小的网络的协议，也是唯一能够妥善处理好不相关路由域间的多路连接的协议。BGP 构建在 EGP 的经验之上。BGP 系统的主要功能是和其他的 BGP 系统交换网络可达信息。网络可达信息包括列出的自治系统（AS）的信息。这些信息有效地构造了 AS 互联的拓扑图并由此清除了路由环路，同时在 AS 级别上可实施策略决策。

实验目的

掌握 BGP 的工作原理，掌握 BGP 的配置。

实验步骤

1. 网络拓扑设计

本实验需要三个 H3C MSR36-20 路由器。图 22-1 所示的 R1、R2 是一个自治域，R1 与 R2 之间使用配置 IBGP，R2 与 R3 之间配置 EBGP，最后通过 R1 ping 通 R3 来检验配置是否正确。

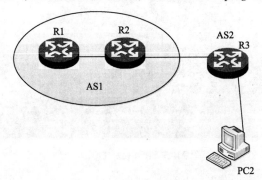

图 22-1　网络拓扑设计

2. 网络地址

具体网络地址规划如表 22-1 所示。

表 22-1　网络地址规划

设备	接口	IP 地址	网关
R1	G0/0	9.1.1.1/24	无
R2	G0/0	9.1.1.2/24	无

续表

设　备	接　口	IP 地　址	网　关
R2	G0/1	3.1.1.1/24	无
R3	G0/0	3.1.1.2/24	无
R3	G0/1	8.1.1.1/24	无
PC2	无	8.1.1.2/24	8.1.1.1

3. 网络配置

R1 的配置为：
```
//进入配置模式
sys
//配置端口
interface g0/0
ip address 9.1.1.1 24
quit
interface loopback 0
ip address 3.3.3.3 32
quit
//配置bgp
bgp 65009
router-id 3.3.3.3
peer 2.2.2.2 as-number 65009
peer 2.2.2.2 connect-interface loopback 0
address-family ipv4 unicast
peer 2.2.2.2 enable
quit
quit
//配置ospf
ospf 1
area 0.0.0.0
network 3.3.3.3 0.0.0.0
network 9.1.1.0 0.0.0.255
quit
quit
```

R3 的配置为：
```
//进入配置模式
sys
//配置端口
interface g0/0
ip address 3.1.1.2 24
quit
interface loopback 0
ip address 1.1.1.1 32
quit
interface g0/1
ip address 8.1.1.1 24
quit
//配置bgp
bgp 65008
router-id 1.1.1.1
peer 3.1.1.1 as-number 65009
```

```
address-family ipv4 unicast
peer 3.1.1.1 enable
network 8.1.1.0 255.255.255.0
quit
quit
```

R2 的配置为：

```
//进入配置模式
sys
//配置端口
interface loopback 0
ip address 2.2.2.2
quit
interface g0/0
ip address 9.1.1.2 24
quit
interface g0/1
ip address 3.1.1.1
quit
//配置 bgp
bgp 65009
router-id 1.1.1.1
peer 3.3.3.3 as-number 65009
Peer 3.1.1.2 as-number 65008
Peer 3.3.3.3 conect-interface loopback 0
address-family ipv4 unicast
network 3.1.1.0 255.255.255.0
network 9.1.1.0 255.255.255.0
peer 3.1.1.2 enable
peer 3.3.3.3 enable
quit
quit
//配置 ospf
ospf 1
area 0.0.0.0
network 2.2.2.2 0.0.0.0
network 9.1.1.0 0.0.0.255
quit
quit
```

实验验证

R1 ping R3 的结果如图 22-2 所示。

```
<R0>ping 8.1.1.1
Ping 8.1.1.1 (8.1.1.1): 56 data bytes, press CTRL_C to break
56 bytes from 8.1.1.1: icmp_seq=0 ttl=254 time=0.486 ms
56 bytes from 8.1.1.1: icmp_seq=1 ttl=254 time=0.195 ms
56 bytes from 8.1.1.1: icmp_seq=2 ttl=254 time=0.181 ms
56 bytes from 8.1.1.1: icmp_seq=3 ttl=254 time=0.197 ms
56 bytes from 8.1.1.1: icmp_seq=4 ttl=254 time=0.169 ms

--- Ping statistics for 8.1.1.1 ---
5 packets transmitted, 5 packets received, 0.0% packet loss
round-trip min/avg/max/std-dev = 0.169/0.246/0.486/0.121 ms
```

图 22-2　R1 ping R3 结果

思考题

如果 R1 属于 AS 65009，R2 和 R3 属于 AS 65008，请重新配置网络，使得 R1、R2、R3 之间可以互相 ping 通。

ACL 配置实验（思科模拟器）

背景介绍

访问控制列表（Access Control List，ACL）是路由器和交换机接口的指令列表，用来控制端口进出的数据包。ACL 适用于所有的路由协议，如 IP、IPX、AppleTalk 等。

信息点间通信和内外网络的通信都是企业网络中必不可少的业务需求，为了保证内网的安全性，需要通过安全策略来保障非授权用户只能访问特定的网络资源，从而达到对访问进行控制的目的。简而言之，ACL 可以过滤网络中的流量，是控制访问的一种网络技术手段。

ACL 是物联网中保障系统安全性的重要技术，在设备硬件层安全基础上，通过对在软件层面对设备间通信进行访问控制，使用可编程方法指定访问规则，防止非法设备破坏系统安全，非法获取系统数据。

实验目的

掌握 ACL 访问控制列表的基本设置方法，了解 ACL 的基本原理。本实验通过思科模拟器实现。

实验步骤

1. 网络拓扑设计

网络拓扑设计如图 23-1 所示。

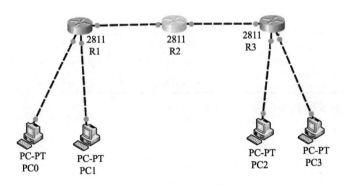

图 23-1　网络拓扑设计

实验所需要的设备为三台 2811 路由器，四台 PC。其中路由器 R2 因为端口不够需要新加板卡。双击 R1 和 R3 路由器打开物理设备视图就可以添加板卡，如图 23-2 所示。

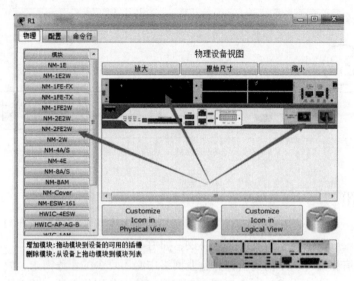

图 23-2 物理界面

2. 网络和服务器配置

各个设备及端口的 IP 地址分配如表 23-1 所示。

表 23-1 IP 地址分配

设 备	接 口	IP 地 址	掩 码	网 关
R1	FastEthernet0/0	192.168.3.1	255.255.255.0	无
R1	FastEthernet0/1	192.168.1.1	255.255.255.0	无
R1	FastEthernet1/0	192.168.2.1	255.255.255.0	无
R2	FastEthernet0/0	192.168.3.2	255.255.255.0	无
R2	FastEthernet0/1	192.168.4.1	255.255.255.0	无
R3	FastEthernet0/0	192.168.4.2	255.255.255.0	无
R3	FastEthernet0/1	192.168.5.1	255.255.255.0	无
R3	FastEthernet1/0	192.168.6.1	255.255.255.0	无
PC0	无	192.168.1.2	255.255.255.0	192.168.1.1
PC1	无	192.168.2.2	255.255.255.0	192.168.2.1
PC2	无	192.168.5.2	255.255.255.0	192.168.5.1
PC3	无	192.168.6.2	255.255.255.0	192.168.6.1

路由器 R1 需要的配置为：
```
//进入特权查看模式
enable
//进入配置模式
config ter
hostname R1
//配置 dhcp 地址池
ip dhcp pool 1
network 192.168.1.0 255.255.255.0
default-router 192.168.1.1
exit
```

```
ip dhcp pool 2
network 192.168.2.0 255.255.255.0
default-router 192.168.2.1
exit
//配置端口
interface fastEthernet0/0
ip address 192.168.3.1 255.255.255.0
no shut
interface fastEthernet0/1
ip address 192.168.1.1 255.255.255.0
no shut
interface fastEthernet1/0
ip address 192.168.2.1 255.255.255.0
no shut
exit
//配置eigrp
router eigrp 1
network 192.168.1.0
network 192.168.2.0
network 192.168.3.0
no auto-summary
```
路由器R2需要的配置为：
```
//进入特权模式
enable
//进入配置模式
config ter
hostname R2
//配置端口
interface loopback0
ip address 2.2.2.2 255.255.255.255
exit
interface fastEthernet0/0
ip address 192.168.3.2 255.255.255.0
ip access-group 1 in
no shut
interface fastEthernet0/1
ip address 192.168.4.1 255.255.255.0
no shut
exit
//配置eigrp
router eigrp 1
network 192.168.3.0
network 192.168.4.0
network 2.0.0.0
no auto-summary
exit
//ACL规则
access-list 1 deny 192.168.2.0 0.0.0.255
access-list 1 permit any
access-list 2 permit 192.168.6.0 0.0.0.255
```

路由器 R3 需要的配置为：
```
//进入特权模式
enable
//进入配置模式
config ter
hostname R1
//配置 dhcp 地址池
ip dhcp pool 1
network 192.168.5.0 255.255.255.0
default-router 192.168.5.1
exit
ip dhcp pool 2
network 192.168.6.0 255.255.255.0
default-router 192.168.6.1
exit
//配置端口
interface fastEthernet0/0
ip address 192.168.4.2 255.255.255.0
no shut
interface fastEthernet0/1
ip address 192.168.5.1 255.255.255.0
no shut
interface fastEthernet1/0
ip address 192.168.6.1 255.255.255.0
no shut
exit
//配置 eigrp
router eigrp 1
network 192.168.4.0
network 192.168.5.0
network 192.168.6.0
no auto-summary
```

实验验证

验证 ACL，因为只在路由器 R2 上设置 ACL：拒绝 PC1 所在的网段访问路由器 R2，则 PC1 不能 ping 通 R2，如图 23-3 所示。

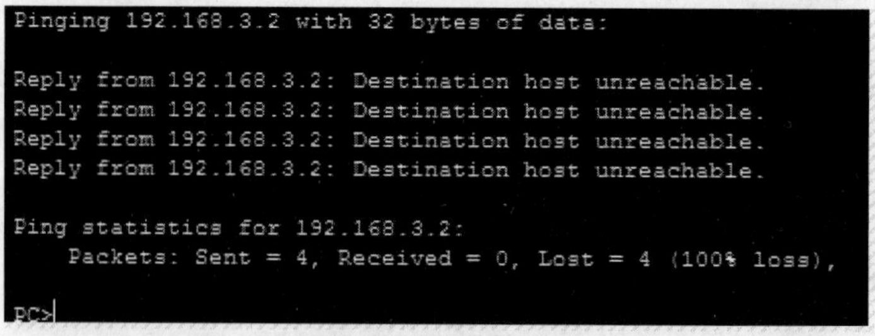

```
Pinging 192.168.3.2 with 32 bytes of data:

Reply from 192.168.3.2: Destination host unreachable.
Reply from 192.168.3.2: Destination host unreachable.
Reply from 192.168.3.2: Destination host unreachable.
Reply from 192.168.3.2: Destination host unreachable.

Ping statistics for 192.168.3.2:
    Packets: Sent = 4, Received = 0, Lost = 4 (100% loss),
PC>
```

图 23-3　PC1 ping R2

其他网段都能 ping 通，验证路由表，保证所有网络的连通性，则 PC0 可以 ping 通 PC2，如图 23-4 所示。

实验 23 ACL 配置实验（思科模拟器） | 79

图 23-4 PC0 ping PC2

PC0 可以 ping 通 PC1，如图 23-5 所示。

图 23-5 PC0 ping PC1

思考题

如果在路由器 R2 上配置并运行远程协议 Telnet，而且只允许 PC2 远程访问路由器 R2 上的 Telnet 服务，该怎么设置？

实验 24 ACL 配置实验（上机）

背景介绍

访问控制列表（ACL）是应用在路由器接口的指令列表（即规则）。这些指令列表用来告诉路由器，哪些数据包可以接受，哪些数据包需要拒绝。

访问控制列表（ACL）的工作原理：

ACL 使用包过滤技术，在路由器上读取 OSI 七层模型的第三层和第四层包头中的信息，如源地址、目标地址、源端口、目标端口等，根据预先定义好的规则，对包进行过滤，从而达到访问控制的目的。

ACL 是一组规则的集合，它应用在路由器的某个接口上。对路由器接口而言，访问控制列表有以下两个方向。

出：已经过路由器的处理，正离开路由器的数据包。

入：已到达路由器接口的数据包，将被路由器处理。

实验目的

掌握 ACL 的配置，了解 ACL 的工作原理。

实验步骤

1. 网络拓扑设计

本实验用到两个交换机都是 H3C s3100v2。PC1 与 PC2 的 IP 配置在同一个网段，PC1 与 PC2 先互 ping 一下，查看结果，然后在两个交换机上配置 ACL 规则，再互 ping 一下，查看结果。网络拓扑设计如图 24-1 所示。

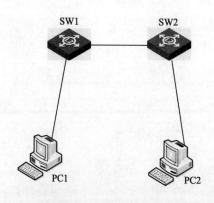

图 24-1　网络拓扑设计

2. 网络地址

具体地址规划如表 24-1 所示。

表 24-1　网络地址规划

设　备	IP　地　址	网　关
PC1	192.168.0.5/24	无
PC2	192.168.0.1/24	无

3. 具体配置

PC1、PC2 只需配置 IP 地址即可。
SW1 的配置为：
```
//进入配置模式
sys
//配置 acl
acl number 2001 name test
rule 5 deny source 192.168.0.1 0
//配置端口
interface eth0/0
port link-type trunk
port trunk permit vlan all
packet-filter 2001 inbound
quit
//划分 vlan
vlan 100
quit
interface etn0/1
port link-type access
port access vlan 100
quit
```
SW2 的配置为：
```
//进入配置模式
sys
//配置 acl
acl number 2001 name test
rule 1 deny source 192.168.0.5 0
//配置端口
interface eth0/0
port link-type trunk
port trunk permit vlan all
packet-filter 2001 inbound
quit
//划分 vlan
vlan 100
quit
interface etn0/1
port link-type access
port access vlan 100
quit
```

 实验验证

没有配置 ACL 之前，如图 24-2 所示。

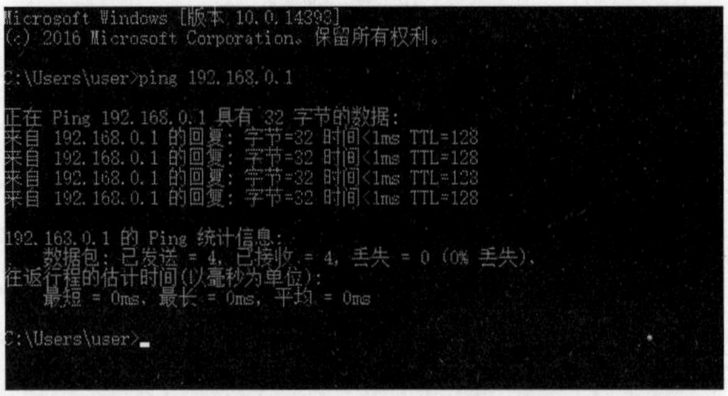

图 24-2　没有配置 ACL 之前

配置了 ACL 之后，如图 24-3 所示。

图 24-3　配置了 ACL 之后

思考题

请删除 ACL 规则，再查看实验结果。

实验 25

→ Telnet 配置实验（思科模拟器）

背景介绍

Telnet 协议是 TCP/IP 协议族中的一员，是 Internet 远程登录服务的标准协议和主要方式。它为用户提供了在本地计算机上完成远程主机工作的能力。在终端使用者的计算机上使用 Telnet 程序，用它连接到服务器。终端使用者可以在 Telnet 程序中输入命令，这些命令会在服务器上运行，就像直接在服务器的控制台上输入一样，可以在本地控制服务器。要开始一个 Telnet 会话，必须输入用户名和密码来登录服务器。Telnet 是常用的远程控制 Web 服务器的方法。

实验目的

掌握 Telnet 的基本设置方法，了解 Telnet 的基本原理。本实验通过思科模拟器实现。

实验步骤

1. 网络拓扑设计

实验所需要的设备为一台 1841 路由器，一台 PC。网络拓扑设计如图 25-1 所示。

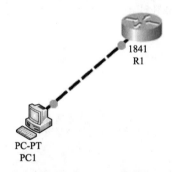

图 25-1　网络拓扑设计

2. 网络和服务器配置

各个设备及端口的 IP 地址分配如表 25-1 所示。

表 25-1　IP 地址分配

设备	接口	IP 地址	掩码	网关
R1	FastEthernet0/0	192.168.1.1	255.255.255.0	无
PC1	无	192.168.1.2	255.255.255.0	192.168.1.1

路由器 R1 需要的配置为：
//进入特权模式
enable
//进入配置
config ter
hostname R1
//配置端口
interface fastethernet0/0
ip address 192.168.1.1 255.255.255.0
no shut
exit
enable password "cisco"
line vty 0 4
password "cisco"
login

实验验证

在 PC 上执行命令 telnet 192.168.1.1（此处登录密码设置为"cisco"），如图 25-2 所示。

图 25-2　telnet 192.168.1.1

如果本实验加入 ACL 访问控制列表，只允许某一台指定的 PC 来远程访问路由，该怎么设置？

实验 26 IPv6 静态路由配置实验（上机）

背景介绍

IPv6 是 Internet Protocol version 6 的缩写，其中 Internet Protocol 译为"互联网协议"。IPv6 是 IETF（互联网工程任务组，Internet Engineering Task Force）设计的用于替代现行版本 IP 协议（IPv4）的下一代 IP 协议。

由于 IPv4 最大的问题在于网络地址资源有限，严重制约了互联网的应用和发展。IPv6 的使用，不仅能解决网络地址资源数量的问题，而且也解决了多种接入设备连入互联网的障碍。

实验目的

掌握 IPv6 的配置，掌握静态路由的配置。

实验步骤

1. 网络拓扑设计

本实验需要两台 H3C MSR36-20 路由器。R1 与 R2 之间配置静态路由和端口，检验时，PC1 能 ping 通 PC2，PC2 能 ping 通 PC1。网络拓扑设计如图 26-1 所示。

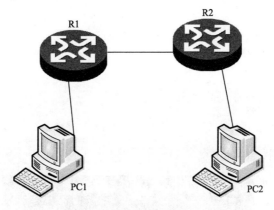

图 26-1 网络拓扑设计

2. 网络地址

网络地址规划如表 26-1 所示。

表 26-1　网络地址规划

设　备	接　口	IP 地　址	网　关
PC1	无	2::2/64	2::1
PC2	无	1::2/64	1::1
R1	G0/0	2::1/64	无
R1	G0/1	3::1/64	无
R2	G0/1	3::2/64	无
R2	G0/0	1::1/64	无

3. 具体网络配置

PC1、PC2 配置 IP 地址即可。

R1 的配置为：

```
//进入配置模式
sys
//配置端口
interface g0/0
ipv6 address 2::1/64
quit
interface g0/1
ipv6 address 3::1/64
quit
//配置静态默认路由
ipv6 route-static 1;: 64 3::2
```

R2 的配置为：

```
//进入配置模式
sys
//配置端口
interface g0/0
ipv6 address 1::1/64
quit
interface g0/1
ipv6 address 3::2/64
quit
//配置静态默认路由
ipv6 route-static 2;: 64 3::1
```

PC2 ping PC1，如图 26-2 所示。

图 26-2　PC2 ping PC1

PC1 ping PC2，如图 26-3 所示。

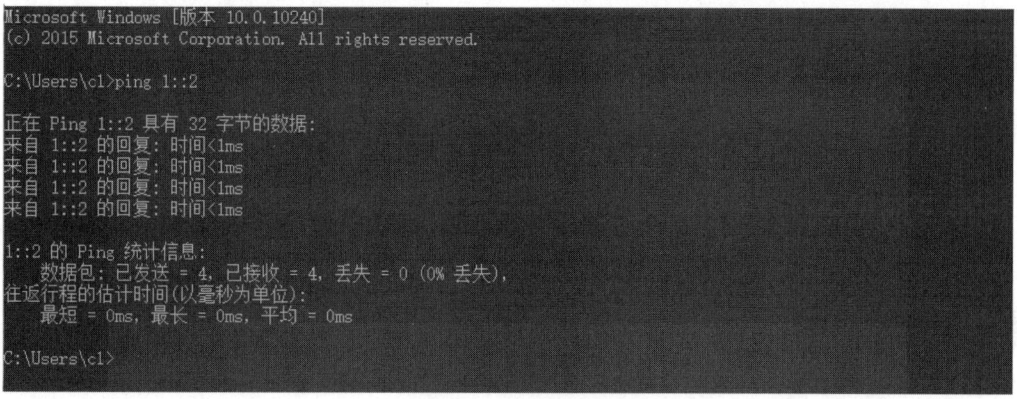

图 26-3　PC1 ping PC2

思考题

请让两台路由器之间用 OSPF 协议实现基于 IPv6 网络。

实验 27

➡ NAT 配置实验（思科模拟器）

背景介绍

NAT（Network Address Translation，网络地址转换）是 1994 年提出的。当专用网内部的一些主机本来已经分配到了本地 IP 地址（即仅在本专用网内使用的专用地址），但现在又想和因特网上的主机通信（并不需要加密）时，可使用 NAT 方法。

这种方法需要在专用网连接到因特网的路由器上安装 NAT 软件。装有 NAT 软件的路由器称为 NAT 路由器，它至少有一个有效的外部全球 IP 地址。这样，所有使用本地地址的主机在和外界通信时，都要在 NAT 路由器上将其本地地址转换成全球 IP 地址，才能和因特网连接。

另外，这种通过使用少量的公有 IP 地址代表较多的私有 IP 地址的方式，将有助于减缓可用的 IP 地址空间的枯竭。在 RFC 1632 中有对 NAT 的说明。

实验目的

掌握 NAT 的基本设置方法，了解 NAT 的基本原理。本次实验通过思科模拟器实现。

实验步骤

1. 网络拓扑设计

实验所需要的设备为两台 1841 路由器，六台 PC，还有一台 Web 服务器。PC 地址是 DHCP 自动分发地址。网络拓扑设计如图 27-1 所示。

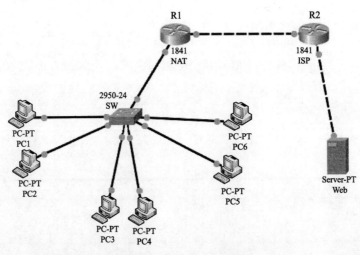

图 27-1　网络拓扑设计

2. 网络和服务器配置

各个设备及端口的 IP 地址分配如表 27-1 所示。

表 27-1 IP 地址分配

设备	接口	IP 地址	掩码	网关	VLAN 号
R1	FastEthernet0/0 .100 FastEthernet0/0 .200 FastEthernet0/0 .300	10.0.0.1 10.0.1.1 10.0.2.1	255.255.255.0 255.255.255.0 255.255.255.0	无	无
R1	FastEthernet0/1	40.0.0.2	255.255.255.0	无	无
R2	FastEthernet0/0	20.0.0.1	255.255.255.0	无	无
R2	FastEthernet0/1	40.0.0.1	255.255.255.0	无	无
PC1	无	10.0.0.2	255.255.255.0	10.0.0.1	100
PC2	无	10.0.0.3	255.255.255.0	10.0.0.1	100
PC3	无	10.0.1.2	255.255.255.0	10.0.1.1	200
PC4	无	10.0.1.3	255.255.255.0	10.0.1.1	200
PC5	无	10.0.2.2	255.255.255.0	10.0.2.1	300
PC6	无	10.0.2.3	255.255.255.0	10.0.2.1	300
Web 服务器	无	20.0.0.2	255.255.255.0	20.0.0.1	无

NAT 路由器和 ISP 路由器通过静态路由协议通信。

路由器 NAT 需要的配置为：

```
//进入特权模式
enable
//进入配置模式
config ter
hostname nat
//配置 dhcp 地址池
ip dhcp pool a
network 10.0.0.0 255.255.255.0
default-router 10.0.0.1
exit
ip dhcp pool b
network 10.0.1.0 255.255.255.0
default-router 10.0.1.1
exit
ip dhcp pool c
network 10.0.2.0 255.255.255.0
default-router 10.0.2.1
exit
//配置子接口
interface fastethernet0/0.100
encapsulation dot1Q 100
ip address 10.0.0.1 255.255.255.0
no shut
ip nat inside
exit
interface fastethernet0/0.200
encapsulation dot1Q 200
```

```
ip address 10.0.1.1 255.255.255.0
no shut
ip nat inside
exit
interface fastethernet0/0.300
encapsulation dot1Q 300
ip address 10.0.2.1 255.255.255.0
no shut
ip nat inside
exit
//配置上联口
interface fastethernet0/1
description to-ISP
ip address 202.102.216.1 255.255.255.252
no shut
ip nat outside
exit
//配置 nat 和默认静态路由
ip nat inside source list 30 interface fastethernet0/1 overload
ip classless
ip route 0.0.0.0 0.0.0.0 fastethernet0/1
access-list 30 permit 10.0.0.0 0.0.0.255
access-list 30 permit 10.0.1.0 0.0.0.255
access-list 30 permit 10.0.2.0 0.0.0.255
end
```
路由器 ISP 需要的配置为:
```
//进入特权模式
enable
//进入配置模式
config ter
hostname isp
//配置端口
interface fastethernet0/0
ip address 202.102.216.2 255.255.255.252
no shut
exit
interface fastethernet0/1
description to-web
ip address 20.0.0.1 255.255.255.0
no shut
exit
```
交换机 SW 需要的配置为:
```
//进入特权模式
enable
//进入配置模式
config ter
hostname sw
//配置中继和接入端口
interface fastethernet0/1
switchport mode trunk
interface fastethernet0/2
switchport access vlan 100
```

```
interface fastethernet0/3
switchport access vlan 100
interface fastethernet0/4
switchport access vlan 200
interface fastethernet0/5
switchport access vlan 200
interface fastethernet0/6
switchport access vlan 300
interface fastethernet0/7
switchport access vlan 300
end
```

实验验证

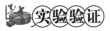

从 PC1 ping PC3、PC5，PC1 ping Web 服务器 IP 地址正常，如图 27-2～图 27-4 所示。

```
PC>ping 10.0.1.2

Pinging 10.0.1.2 with 32 bytes of data:

Reply from 10.0.1.2: bytes=32 time=0ms TTL=127
Reply from 10.0.1.2: bytes=32 time=0ms TTL=127
Reply from 10.0.1.2: bytes=32 time=0ms TTL=127
Reply from 10.0.1.2: bytes=32 time=0ms TTL=127

Ping statistics for 10.0.1.2:
    Packets: Sent = 4, Received = 4, Lost = 0 (0% loss),
Approximate round trip times in milli-seconds:
    Minimum = 0ms, Maximum = 0ms, Average = 0ms
```

图 27-2　PC1 ping PC3

```
PC>ping 10.0.2.2

Pinging 10.0.2.2 with 32 bytes of data:

Request timed out.
Reply from 10.0.2.2: bytes=32 time=1ms TTL=127
Reply from 10.0.2.2: bytes=32 time=0ms TTL=127
Reply from 10.0.2.2: bytes=32 time=0ms TTL=127

Ping statistics for 10.0.2.2:
    Packets: Sent = 4, Received = 3, Lost = 1 (25% loss),
Approximate round trip times in milli-seconds:
    Minimum = 0ms, Maximum = 1ms, Average = 0ms
```

图 27-3　PC1 ping PC5

```
PC>ping 20.0.0.2

Pinging 20.0.0.2 with 32 bytes of data:

Reply from 20.0.0.2: bytes=32 time=12ms TTL=126
Reply from 20.0.0.2: bytes=32 time=1ms TTL=126
Reply from 20.0.0.2: bytes=32 time=0ms TTL=126
Reply from 20.0.0.2: bytes=32 time=0ms TTL=126

Ping statistics for 20.0.0.2:
    Packets: Sent = 4, Received = 4, Lost = 0 (0% loss),
Approximate round trip times in milli-seconds:
    Minimum = 0ms, Maximum = 12ms, Average = 3ms
```

图 27-4　PC1 ping Web 服务器

思考题

如果让 NAT 和 OSPF，或是 NAT 和 RIP 在一起配置路由，那么 NAT 该怎么融入 OSPF 或 RIP 中呢？

实验 28 NAT 配置实验（上机）

背景介绍

NAT 英文全称是"Network Address Translation"，中文含义是"网络地址转换"，它是一个 IETF（Internet Engineering Task Force，Internet 工程任务组）标准，允许一个整体机构以一个公用 IP（Internet Protocol）地址出现在 Internet 上。顾名思义，它是一种把内部私有网络地址（IP 地址）翻译成合法网络 IP 地址的技术。

实验目的

了解 NAT 的实验原理，掌握 NAT 的配置。

实验步骤

1. 网络拓扑设计

实验用了一台 H3C MSR36-20 路由器和 H3C Secpath F100-C-SI 防火墙。图 28-1 所示的防火墙需要配置 NAT 和默认静态路由，R1 只需配置端口和 loopback 地址，PC1 也只需要配置 IP 地址，PC2 使用 DHCP 动态获取 IP 地址，检验时，用 PC2 ping loopback 地址，然后在防火墙上查看地址翻译情况。

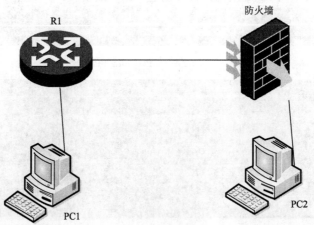

图 28-1　网络拓扑设计

2. 网络地址规划

具体网络地址规划如表 28-1 所示。

表 28-1 网络地址规划

设 备	接 口	IP 地 址	网 关
防火墙	Eth0/0	192.168.0.1/24	无
防火墙	Eth0/1	202.102.216.2/30	无
R1	G0/0	172.16.110.1/24	无
R1	G0/1	202.102.216.1/30	无

3. 具体网络配置

防火墙的配置为：
//进入配置模式
sys
//配置dhcp和nat
dhcp enable
dhcp service ip pool a
network 192.168.0.0 255.255.255.0
gateway-list 192.168.0.1
quit
nat address-group 1
network 202.102.216.2 202.102.216.2
quit
acl number 2000
rule 1 permit source 192.168.0.0 0.0.0.255
quit
//配置端口
interface eth0/0
ip address 192.168.0.1 255.255.255.0
quit
interface eth0/1
ip address 202.102.216.2 255.255.255.252
nat outbound address-group 1
quit
//配置默认静态路由
ip route-static 0.0.0.0 0.0.0.0 eth0/1 202.102.216.1
quit
ip ttl-expires enable
ip unreachable enable
R1 的配置为：
//进入配置模式
sys
//配置端口
interface g0/0
ip address 172.16.110.1 255.255.255.0
quit
//配置端口
interface g0/1

```
ip address 202.102.216.1 255.255.255.252
quit
interface loopback 0
ip address 1.1.1.1 255.255.255.255
ip ttl-expires enable
ip unreachable enable
quit
```

实验验证

用 PC2 ping R1 在防火墙上看到的翻译如图 28-2 所示。

```
[NAT]dis nat session
There are currently 8 NAT sessions:

Pro    GlobalAddr:Port           LocalAddr:Port         DestAddr:Port
ICMP   202.102.216.2:12300       192.168.0.2:1          172.16.110.2:1
       GlobalVPN: ---            LocalVPN: ---
       status: 1                 TTL: 00:00:10          Left: 00:00:10
```

图 28-2　PC2 ping R1

思考题

如果再增加一个路由器，或者三层交换机该怎样配置？

TCP socket 编程

背景介绍

TCP/IP（Transmission Control Protocol/Internet Protocol）即传输控制协议/网间协议，是一种面向连接（连接导向）的、可靠的、基于字节流的运输层（Transport layer）通信协议，由 IETF 的 RFC 793 说明（specified）。在简化的计算机网络 OSI 模型中，它完成第四层传输层所指定的功能，UDP 是同一层内另一个重要的传输协议。

实验目的

两台计算机通信，那么数据是如何在两台计算机之间传输的呢？通过网线来传输，通过电的正负来表示二进制中的 0 和 1，大家都知道在计算机中，一切数据（图像、声音、文字等）都是二进制方式存储的，所以这样便可以通过网线传输任何数据。

那究竟如何编程在传输层实现数据的传输呢？难道用高级程序语言控制电的正负极吗？当然不需要这样做，只需要使用操作系统提供的一套网络编程的 API 函数即可。什么是 API 函数？API 其实也就是函数，只要学会使用这些函数，那么就可以用来编写出各式各样的网络程序，不需要去做重复造轮子的事情。

本次实验的目的主要有以下三点：掌握 socket 管道的基本原理；掌握通过 socket 编程实现 C/S（客户端/服务器端）程序的基本方法；掌握利用 Windows socket 函数库在 WIN 32 平台下编制通信程序的方法。

实验步骤

程序分为服务端和客户端，服务端就相当于网站服务器，客户端就相当于浏览器。

在服务端，需要按照以下步骤：

（1）初始化 socket 库。
（2）绑定本机地址和端口（服务端特有）。
（3）监听端口，等待客户端连接。
（4）当有客户端连接，进行处理，继续监听或者结束程序。
（5）退出程序，关闭 socket，终止对 socket 库的使用。

在客户端，需要按照以下步骤：

（1）初始化 socket 库。
（2）设置远程主机的地址和端口信息，并连接。
（3）等待服务端的响应。
（4）当服务端响应，进行处理。

（5）退出程序，关闭 socket，终止对 socket 库的使用。

图 29-1 所示已直观地说明通信的过程。

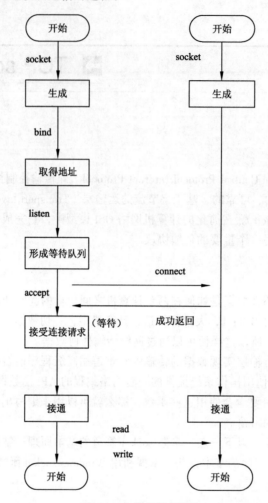

图 29-1　通信的过程

代码示例如下：

服务端：

```
#include <stdio.h>
#include <winsock2.h>
#pragma comment(lib, "ws2_32.lib") //引入winsock库
int main(void)
{
    int len = 0;
    wsadata wd;
    int ret = 0;
    socket s,c;
    char sendbuf[1000]="", recvbuf[1000]="";
    sockaddr_in saddr, caddr;
    ret = wsastartup(makeword(2,2),&wd);
    /*1.初始化操作*/
    if(ret != 0)
```

```c
    {
        return 0;
    }
    if(hibyte (wd.wversion)!=2 || lobyte(wd.wversion)!=2)
    {
        printf("初始化失败");
        wsacleanup();
        return 1;
    }
    /*2.创建服务端socket*/
    s = socket(af_inet, sock_stream, 0);
    /*3.设置服务端信息*/
    saddr.sin_addr.S_un.S_addr = htonl(inaddr_any);   //可以设置参数为IP地址
    saddr.sin_family = af_inet; /*协议类型*/
    saddr.sin_port = htons(8888);
    /*4.绑定在本地端口*/
    bind(s, (sockaddr *)&saddr, sizeof(sockaddr));
    /*5.监听端口*/
    listen(s,5);
    len = sizeof(sockaddr);
    while(1)
    {
        /*6.等待客户端连接，会阻塞在此处，直到有客户端连接到来。*/
        c = accept(s, (sockaddr*)&caddr, &len);
        //格式化字符串
        sprintf(sendBuf,"欢迎登陆服务器，您的ip地址为: %s\n", inet_ntoa(caddr.sin_addr));
        /*7.发送数据到客户端*/
        send(c, sendBuf, strlen(sendBuf)+1, 0);
        /*8.接受客户端的返回*/
        recv(c, recvBuf, 1000, 0);
        /*9.打印出客户端发送来的数据*/
        printf("%s\n", recvBuf);
        /*10.如果不再跟这个客户端联系，就关闭它*/
        closesocket(c);
    }
    /*如果有退出循环的条件，这里还需要清除对socket库的使用*/
    /* wsacleanup();*/
    return 0;
}
```

客户端：
```c
#include <stdio.h>
#include <winsock2.h>
#pragma comment(lib, "ws2_32.lib")  //引入winsock库
int main(void)
{
    wsadata  wd;
    int ret = 0;
    socket c;
    char recvbuf[1000]="", sendbuf[1000]="";
    sockaddr_in saddr;
```

```
ret = wsastartup(makeword(2,2),&wd);
/*1.初始化操作*/
if(ret != 0)
{
    return 0;
}
if(hibyte(wd.wVersion)!=2 || lobyte(wd.wversion)!=2)
{
    printf("初始化失败");
    wsacleanup();
    return 1;
}
/*2.创建客户端socket*/
c = socket(af_inet, sock_stream, 0);
/*3.定义要连接的服务端信息*/
saddr.sin_addr.S_un.S_addr = inet_addr("127.0.0.1");
saddr.sin_family = af_inet;
saddr.sin_port = htons(8888);
/*4.连接服务端*/
connect(c, (sockaddr*)&saddr, sizeof(sockaddr));
recv(c, recvBuf, 1000, 0);
printf("服务端发来的数据:%s\n", recvBuf);
sprintf(sendBuf, "客户端发来的数据: 服务端你好! ");
send(c, sendBuf, strlen(sendBuf)+1, 0);
closesocket(c);
wsacleanup();
return 0;
}
```

实验验证

gcc 下的编译命令:

```
gcc -o server.exe server.c -lwsock32
gcc -o client.exe client.c -lwsock32
```

如果是 vc 或者 vs 编译器下,可以在代码中的#include <winsock2.h>下面加上一行 "#pragma comment(lib, "ws2_32.lib") //此处无分号结尾,用于引入 winsock 库"。

cmd 命令行执行:打开两个 cmd,进入相应的目录,分别输入 server.exe 和 client.exe,运行结果如图 29-2 所示。

图 29-2　运行结果

实验 30

→ DNS 和 Web 配置实验（思科模拟器）

📞 背景介绍

DNS（Domain Name System，域名系统），万维网上作为域名和 IP 地址相互映射的一个分布式数据库，能够使用户更方便地访问互联网，而不用去记住能够被机器直接读取的 IP 地址串。通过域名，最终得到该域名对应的 IP 地址的过程称为域名解析（或主机名解析）。

Web（World Wide Web）即全球广域网，又称万维网，它是一种基于超文本和 HTTP 的、全球性的、动态交互的、跨平台的分布式图形信息系统，是建立在 Internet 上的一种网络服务，为浏览者在 Internet 上查找和浏览信息提供了图形化的、易于访问的直观界面，其中的文档及超链接将 Internet 上的信息结点组织成一个互为关联的网状结构。

⏳ 实验目的

掌握 DNS 服务器和 WEB 服务器的基本设置方法，了解 DNS 和 Web 的基本原理。本实验通过思科模拟器实现。

👆 实验步骤

1. 网络拓扑设计

实验所需要的设备为两台 2811 路由器；三台服务器：一台充当 DNS 服务器，一台充当淘宝 Web 服务器，一台充当京东 Web 服务器；一台 PC。网络拓扑设计如图 30-1 所示。

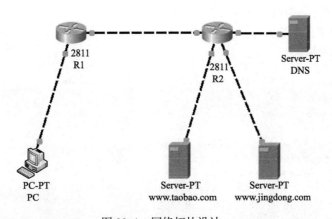

图 30-1 网络拓扑设计

其中路由器 R2 因为端口不够，需要新加板卡。双击 R2 路由器打开物理设备视图就可以添加板卡，具体如图 30-2 所示。

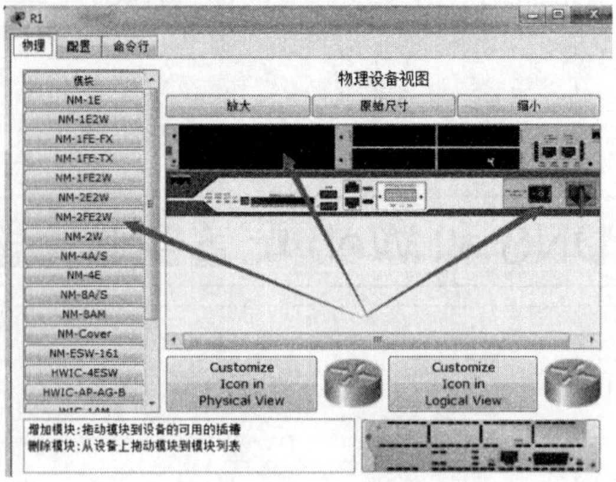

图 30-2　物理界面

2. 网络和服务器配置

各个设备及端口的 IP 地址分配如表 30-1 所示。

表 30-1　IP 地址分配

设备	接口	IP 地址	掩码	网关	DNS
R1	FastEthernet0/0	20.0.0.1	255.255.255.252	无	无
R1	FastEthernet0/1	10.0.0.1	255.255.255.252	无	无
R2	FastEthernet0/0	20.0.0.2	255.255.255.252	无	无
R2	FastEthernet0/1	40.0.0.1	255.255.255.252	无	无
R2	FastEthernet1/0	50.0.0.1	255.255.255.252	无	无
R2	FastEthernet1/1	30.0.0.1	255.255.255.252	无	无
PC	无	10.0.0.2	255.255.255.252	10.0.0.1	30.0.0.2
DNS	无	30.0.0.2	255.255.255.252	30.0.0.1	无
www.taobao.com	无	40.0.0.2	255.255.255.252	40.0.0.1	无
www.jingdong.com	无	50.0.0.2	255.255.255.252	50.0.0.1	无

R1 和 R2 通过静态路由协议通信。

路由器 R1 需要的配置为：

```
//进入特权查看模式
enable
//进入配置模式
conf terminal
//设定路由器名称
hostname R1
//设定端口 fastethernet0/0 信息
interface fastethernet0/0
description to-R2-fa0/0
no shut
ip address 20.0.0.1 255.255.255.252
exit
//设定端口 fastethernet0/1 信息
interface fastethernet0/1
description to-PC
ip address 10.0.0.1 255.255.255.252
```

```
no shut
exit
//设定静态默认路由
ip route 0.0.0.0 0.0.0.0 fastethernet0/0
```
路由器 R2 需要的配置为:
```
//进入特权查看模式
enable
//进入配置模式
conf terminal
//设定路由器名称
hostname R2
//设定端口 fastethernet0/0 信息
interface fastethernet0/0
description to-R2-fa0/0
ip address 20.0.0.2 255.255.255.252
no shut
exit
//设定端口 fastethernet0/1 信息
interface fastethernet0/1
description to-taobao
ip address 40.0.0.1 255.255.255.252
no shut
exit
//设定端口 fastethernet1/0 信息
interface fastethernet1/0
ip address 50.0.0.1 255.255.255.252
no shut
exit
//设定端口 fastethernet1/1 信息
interface fastethernet1/1
description to-DNS
ip address 30.0.0.1 255.255.255.252
no shut
exit
//设置静态路由
ip route 10.0.0.0 255.255.255.252 FastEthernet0/0
```
DNS 服务器配置如图 30-3 所示。

图 30-3 DNS 服务器配置

双击 DNS 服务器，在配置中选择 DNS 标签，添加两个 A Record：一个对应淘宝，一个对应京东。A Record 就是网站和 IP 地址的对应记录，在 Name 编辑框中添加网站网址，在 Address 中添加对应的 IP 地址。具体服务器的网络配置如图 30-4 所示。

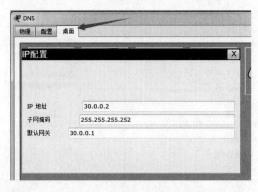

图 30-4　DNS 服务器网络配置

淘宝服务器的 Web 配置如图 30-5 所示。

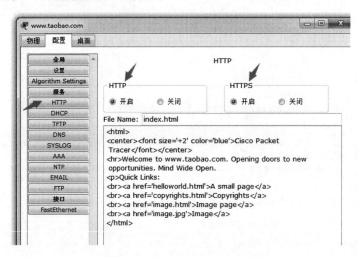

图 30-5　淘宝服务器 Web 配置

双击淘宝服务器，单击 HTTP 标签，开启 HTTP 和 HTTPS 服务，修改主页的欢迎词为：Welcome to www.taobao.com。淘宝服务器的网络设置如图 30-6 所示。

请按同样的方法设置京东服务器。PC 终端只需要设置网络信息，如图 30-7 所示。

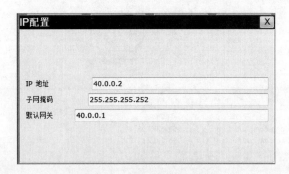

图 30-6　淘宝服务器网络配置

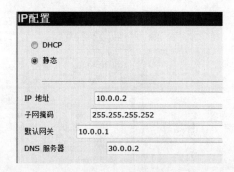

图 30-7　PC 网络配置

实验验证

1. 网络验证

从 PC ping DNS、淘宝和京东服务器 IP 地址正常，如图 30-8 所示。

图 30-8　网络验证

2. 业务验证

从 PC 访问淘宝、京东主页正常，如图 30-9 所示。

图 30-9　业务验证

思考题

如果在网络拓扑图 30-1 中新加一台当当服务期 60.0.0.2/30，此时 DNS 服务器和网络设备需要如何增加配置？

实验 31

→ FTP 配置实验（思科模拟器）

背景介绍

FTP（File Transfer Protocol，文件传输协议）是 TCP/IP 协议组中的协议之一。FTP 协议包括两个组成部分，其一为 FTP 服务器，其二为 FTP 客户端。其中 FTP 服务器用来存储文件，用户可以使用 FTP 客户端通过 FTP 协议访问位于 FTP 服务器上的资源。在开发网站的时候，通常利用 FTP 协议把网页或程序传到 Web 服务器上。此外，由于 FTP 传输效率非常高，在网络上传输大的文件时，一般也采用该协议。

默认情况下，FTP 协议使用 TCP 端口中的 20 和 21 这两个端口，其中 20 用于传输数据，21 用于传输控制信息。但是，是否使用 20 作为传输数据的端口与 FTP 使用的传输模式有关，如果采用主动模式，那么数据传输端口就是 20；如果采用被动模式，则具体最终使用哪个端口需服务器端和客户端协商决定。

实验目的

掌握 FTP 服务器的配置，了解 FTP 服务器的工作原理。本实验通过思科模拟器来实现。

实验步骤

1. 网络拓扑设计

实验需要三台 2811 的路由器，一台 PC，一台 FTP 服务器。三台路由器之间通过默认静态路由进行通信，注意 R2 需要配置返程路由。网络拓扑设计如图 31-1 所示。

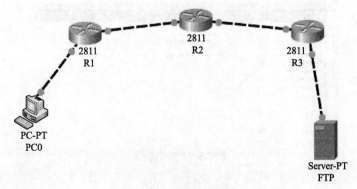

图 31-1 网络拓扑设计

2. 网络地址

具体的网络地址规划如表 31-1 所示。

表 31-1 网络地址规划

设 备	接 口	IP 地 址	网 关
R1	Fa0/0	172.16.100.1/30	无
R1	Fa0/1	192.168.1.1/30	无
R2	Fa0/0	192.168.1.2/30	无
R2	Fa0/1	192.168.2.1/30	无
R3	Fa0/0	192.168.2.2/30	无
R3	Fa0/1	192.168.0.1/30	无
PC0	无	172.16.100.2/30	172.16.100.1
FTP	无	192.168.0.2/30	192.168.0.1

2．具体配置命令

路由器 R1 配置为：
//进入特权查看模式
enable
//进入配置模式
conf terminal
//配置端口
interface fastethernet0/0
description to PC0
ip address 172.16.100.1 255.255.255.252
duplex auto
speed auto
interface fastethernet0/1
description to R2
ip address 192.168.1.1 255.255.255.252
duplex auto
speed auto
interface vlan1
no ip address
shutdown
ip classless
//配置默认静态路由
ip route 0.0.0.0 0.0.0.0 fastethernet0/1
路由器 R2 配置为：
//配置端口
interface fastethernet0/0
description to R1
ip address 192.168.1.2 255.255.255.252
duplex auto
speed auto
interface fastethernet0/1
description to R3
ip address 192.168.2.1 255.255.255.252
duplex auto
speed auto
interface vlan1
no ip address
shutdown
ip classless

```
//配置默认静态路由
ip route 0.0.0.0 0.0.0.0 fastethernet0/1
ip route 0.0.0.0 0.0.0.0 fastethernet0/0
```
路由器 R3 配置：
```
//配置端口
interface fastethernet0/0
description to R2
ip address 192.168.2.2 255.255.255.252
duplex auto
speed auto
interface fastethernet0/1
description to ftp
ip address 192.168.0.1 255.255.255.252
duplex auto
speed auto
interface vlan1
no ip address
shutdown
ip classless
//配置静态默认路由
ip route 0.0.0.0 0.0.0.0 fastethernet0/0
```
PC0 的设置如图 31-2 和图 31-3 所示。

图 31-2　PC0 设置 1

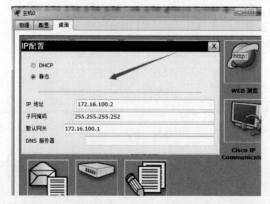

图 31-3　PC0 设置 2

FTP 服务器的设置如图 31-4 和图 31-5 所示。

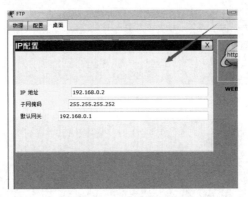

图 31-4　FTP 服务器设置 1

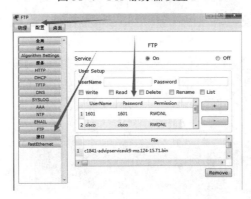

图 31-5　FTP 服务器设置 2

实验验证

连接 FTP 服务器，如图 31-6 ~ 图 31-11 所示。

图 31-6　连接 FTP 服务器 1

图 31-7　连接 FTP 服务器 2

图 31-8 连接 FTP 服务器 3

图 31-9 连接 FTP 服务器 4

图 31-10 连接 FTP 服务器 5

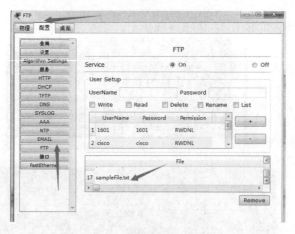

图 31-11 连接 FTP 服务器 6

本实验用的是 RIP 协议，请用 OSPF 协议实现在三台路由器之间转发路由。

实验 32

→ 路由器端口镜像配置（上机）

背景介绍

端口镜像（Port Mirroring）功能通过在交换机或路由器上，将一个或多个源端口的数据流量转发到某一个指定端口来实现对网络的监听，指定端口称之为"镜像端口"或"目的端口"，在不严重影响源端口正常吞吐流量的情况下，可以通过镜像端口对网络的流量进行监控分析。在企业中用镜像功能，可以很好地对企业内部的网络数据进行监控管理，在网络出故障的时候，可以快速定位故障。

实验目的

掌握端口镜像的配置，掌握端口镜像的实验原理。学会使用抓包工具。

实验步骤

1. 网络拓扑设计

本实验用到两台 H3C MSR36-20 路由器。在 R1 G0/1 端口配置目的端口，R2 G0/0 端口配置镜像端口，检验时，通过 R1 ping R2，在终端通过 wireshark 可以看到数据包。网络拓扑设计如图 32-1 所示。

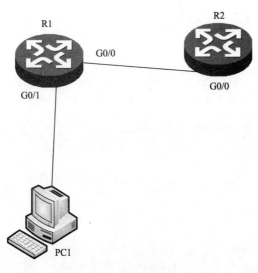

图 32-1　网络拓扑设计

2. 网络地址

具体地址规划,如表 32-1 所示。

表 32-1 网络地址规划

设 备	端 口	IP 地址	网 关
R1	G0/0	202.102.216.2/30	无
R1	G0/1	192.168.2.1/24	无
R2	G0/0	202.102.216.1/30	无
PC1	无	192.168.2.2/24	192.168.2.1

3. 具体的配置

R1 的配置为:
```
//进入配置模式
sys
//配置 loopback 地址
interface loopback0
ip address 1.1.1.1 255.255.255.255
quit
//配置端口
mirroring-group 1 local
interface g0/0
ip address 202.102.216.2 255.255.255.252
port link-mode route
mirroring-group 1 mirroring-port both
quit
//配置抓包端口
mirroring-group 1 local
interface g0/1
ip address 192.168.2.1 255.255.255.0
port link-mode route
mirroring-group 1 monitor-port
quit
```
R2 的配置为:
```
sys
//配置 loopback 地址
interface loopback0
ip address 2.2.2.2 255.255.255.255
quit
//配置端口
interface g0/0
ip address 202.102.216.1 255.255.255.252
port link-mode route
quit
```

 实验验证

通过 R1 ping R2,并打开 wireshark,ping 2.2.2.2 –c 1000,如图 32-2 所示。

```
address 202.102.216.2 202.102.216.2
#
return
[NAT]ping -c 1000 2.2.2.2
Ping 2.2.2.2 (2.2.2.2): 56 data bytes, press CTRL_C to break
56 bytes from 2.2.2.2: icmp_seq=0 ttl=255 time=0.377 ms
56 bytes from 2.2.2.2: icmp_seq=1 ttl=255 time=0.167 ms
56 bytes from 2.2.2.2: icmp_seq=2 ttl=255 time=0.140 ms
56 bytes from 2.2.2.2: icmp_seq=3 ttl=255 time=0.174 ms
56 bytes from 2.2.2.2: icmp_seq=4 ttl=255 time=0.167 ms
56 bytes from 2.2.2.2: icmp_seq=5 ttl=255 time=0.245 ms
56 bytes from 2.2.2.2: icmp_seq=6 ttl=255 time=0.152 ms
56 bytes from 2.2.2.2: icmp_seq=7 ttl=255 time=0.144 ms
56 bytes from 2.2.2.2: icmp_seq=8 ttl=255 time=0.141 ms
56 bytes from 2.2.2.2: icmp_seq=9 ttl=255 time=0.152 ms
56 bytes from 2.2.2.2: icmp_seq=10 ttl=255 time=0.154 ms
56 bytes from 2.2.2.2: icmp_seq=11 ttl=255 time=0.160 ms
```

图 32-2　R1 ping R2

在 wireshark 中观察数据包，如图 32-3 所示。

No.	Time	Source	Destination	Protocol	Info
1	0.000000	169.254.80.235	169.254.255.255	BROWSE	Local Master Announcement DESKTO
2	5.942313	2.2.2.2	202.102.216.2	ICMP	Echo (ping) reply
3	5.942313	202.102.216.2	2.2.2.2	ICMP	Echo (ping) request
4	6.142811	2.2.2.2	202.102.216.2	ICMP	Echo (ping) reply
5	6.142844	202.102.216.2	2.2.2.2	ICMP	Echo (ping) request
6	6.343119	2.2.2.2	202.102.216.2	ICMP	Echo (ping) reply
7	6.343147	202.102.216.2	2.2.2.2	ICMP	Echo (ping) request
8	6.543435	2.2.2.2	202.102.216.2	ICMP	Echo (ping) reply
9	6.543469	202.102.216.2	2.2.2.2	ICMP	Echo (ping) request
10	6.743689	2.2.2.2	202.102.216.2	ICMP	Echo (ping) reply
11	6.743723	202.102.216.2	2.2.2.2	ICMP	Echo (ping) request
12	6.944045	2.2.2.2	202.102.216.2	ICMP	Echo (ping) reply
13	6.944073	202.102.216.2	2.2.2.2	ICMP	Echo (ping) request
14	7.144457	2.2.2.2	202.102.216.2	ICMP	Echo (ping) reply
15	7.144485	202.102.216.2	2.2.2.2	ICMP	Echo (ping) request
16	7.344694	2.2.2.2	202.102.216.2	ICMP	Echo (ping) reply
17	7.344694	202.102.216.2	2.2.2.2	ICMP	Echo (ping) request
18	7.545048	2.2.2.2	202.102.216.2	ICMP	Echo (ping) reply

图 32-3　观察数据包

思考题

请在路由器 R1 上配置 NAT，并捕获查看相关数据包。

实验 33 无线网络设计（思科模拟器）

背景介绍

无线接入点即无线 AP（Access Point），它是一个无线网络的接入点，用于无线网络的无线交换机，也是无线网络的核心，主要在媒体存取控制层 MAC 中扮演无线工作站及有线局域网络的桥梁。无线路由器主要由路由交换接入一体设备和纯接入点设备，一体设备执行接入和路由工作，纯接入设备只负责无线客户端的接入，纯接入设备通常作为无线网络扩展使用，与其他 AP 或者主 AP 连接，以扩大无线覆盖范围。无线接入点主要用于家庭宽带、大楼内部及园区内部，典型距离覆盖几十米至上百米，目前主要技术为 802.11 系列。

实验目的

掌握无线路由器的基本设置方法，了解无线路由器基本原理。本实验通过思科模拟器实现。

实验步骤

1. 网络拓扑设计

网络拓扑设计如图 33-1 所示。

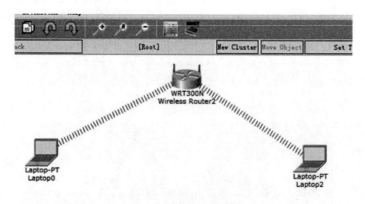

图 33-1　网络拓扑设计

实验所需要的设备为一台 WRT300N 无线路由器，两台笔记本式计算机。具体如图 33-2 所示。

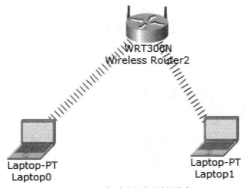

图 33-2　实验所需要的设备

2．设置步骤

（1）打开笔记本式计算机属性对话框，关闭笔记本式计算机电源，为笔记本式计算机安装无线网卡，默认情况下，它是安装有线网卡，安装完毕之后，打开笔记本式计算机电源，此时笔记本式计算机就会自动搜索无线信号，如图 33-3 所示。

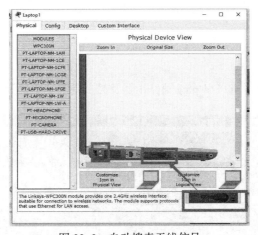

图 33-3　自动搜索无线信号

（2）打开无线路由器属性对话框，开启并设置无线路由器地址范围和用户数，如图 33-4 所示。

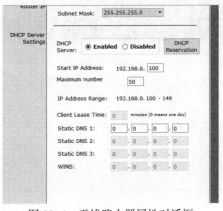

图 33-4　无线路由器属性对话框

（3）设置路由器的登录密码，如图 33-5 所示。保存关闭该窗口。
（4）打开笔记本式计算机，将 IP 配置设置为 DHCP，自动获取 IP 地址，如图 33-6 所示。

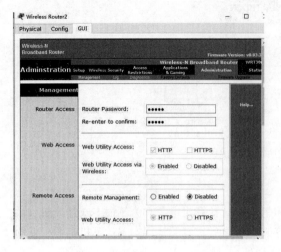
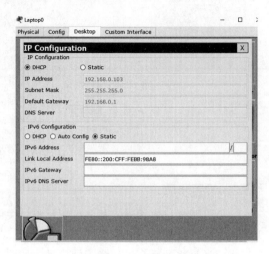

图 33-5　设置路由器的登录密码　　　　　　　　图 33-6　设置为 DHCP

网络验证

其中一台笔记本式计算机可以 ping 通路由器及另一台笔记本式计算机，如图 33-7 所示。

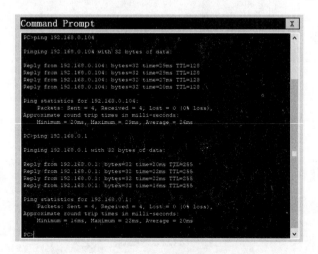

图 33-7　网络验证

无线 AP 和无线路由器的区别是什么？

实验 34 → 无线网络设计（上机）

背景介绍

无线接入点即无线 AP（Access Point），它是一个无线网络的接入点，用于无线网络的无线交换机，也是无线网络的核心，主要在媒体存取控制层 MAC 中扮演无线工作站及有线局域网络的桥梁。无线路由器主要由路由交换接入一体设备和纯接入点设备，一体设备执行接入和路由工作，纯接入设备只负责无线客户端的接入，纯接入设备通常作为无线网络扩展使用，与其他 AP 或者主 AP 连接，以扩大无线覆盖范围。无线接入点主要用于家庭宽带、大楼内部及园区内部，典型距离覆盖几十米至上百米，目前主要技术为 802.11 系列。

实验目的

掌握无线路由器的基本设置方法，了解无线路由器基本原理。本实验通过华为实体设备实现。

实验步骤

网络拓扑设计

网络拓扑设计如图 34-1 所示。

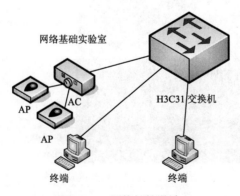

图 34-1 网络拓扑设计

每个实验室放置一台 AC 和两台 AP，AC 上进行 NAT（地址翻译）IP 地址翻译，每台 AC 只需要分配一个 IP 地址，AC 负责把翻译过后的 IP 地址分配给各个无线终端，无线 AP 只负责无线信号发送。

AC 设备的具体关键配置为，先配置逻辑 AC 如下：

```
dhcp server ip-pool ap
network 192.168.8.0 mask 255.255.255.0        //发布dhcp地址
```

```
gateway-list 192.168.8.1                    //设置网关
dns-list 202.102.213.68 61.132.163.68       //设置域名解析地址
interface vlan-interface20                  //设置无线接入终端信息
description ap-gateway
ip address 192.168.8.1 255.255.255.0
interface vlan-interface50                  //设置出口信息
description nat and admin
ip address 172.16.116.9 255.255.255.0
nat outbound 2000 address-group 0
wlan ap-group default_group                 //设置管理的fit ap信息
ap ap1
ap ap2
wlan service-template 1 clear
ssid wl1501
bind wlan-ESS 1
service-template enable
oap connect slot 1 切换到逻辑交换机后的关键配置如下：
interface gigabitethernet1/0/6              //无线ap接入端口配置
port access vlan 20
poe enable                                  //打开poe供电
interface gigabitethernet1/0/2              //ac出口配置
port access vlan 50
poe enable
```

实验验证

网络构建完成后，在实验室测试网络运行正常，无线设备可以正常接入，访问外网也正常，具体如图34-2所示。

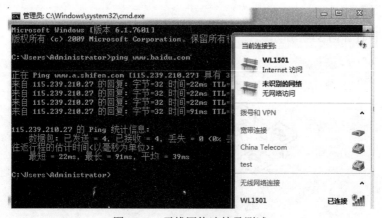

图34-2 无线网络连接及测试

思考题

无线AP和无线路由器的区别是什么？

实验 35 网吧网络设计（思科模拟器）

背景介绍

网吧网络设计运用到 NAT 转换技术，NAT 的实现方式有三种，本实验中运用到的是端口复用 overload，即改变外出数据包的源端口并进行转换。这种方法能让内部网络的所有主机均共享一个合法的外部 IP 地址，可以最大限度地节约 IP 地址资源。同时又能隐藏网络内部的所有主机，有效避免 Internet 的攻击。

实验目的

了解 NAT 转换的基本原理，掌握小型网吧的网络设计及设备的配置方法。

实验步骤

1. 网络拓扑设计

实验所需的设备为四台 PC，三台 2950-24 交换机，两台 1841 路由器，一台服务器充当外网服务器。网络拓扑设计如图 35-1 所示。

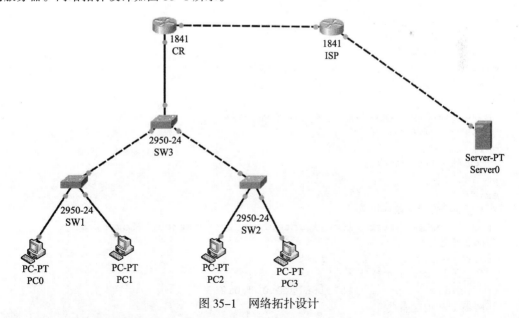

图 35-1 网络拓扑设计

2. 网络和服务器配置

各个设备及端口的 IP 地址分配如表 35-1 所示。

表 35-1　IP 地址分配

设　备	接　口	IP 地址	掩　码	网　关
PC0	Fa0/0	192.168.0.3	255.255.255.128	192.168.0.1
PC1	Fa0/0	192.168.0.131	255.255.255.128	192.168.0.129
PC2	Fa0/0	192.168.0.2	255.255.255.128	192.168.0.1
PC3	Fa0/0	192.168.0.135	255.255.255.128	192.168.0.129
CR	Fa0/1.100	192.168.0.1	255.255.255.128	无
CR	Fa0/1.200	192.168.0.129	255.255.255.128	无
CR	Fa0/0	202.102.216.2	255.255.255.0	无
ISP	Fa0/0	202.102.216.1	255.255.255.0	无
ISP	Fa0/1	61.132.246.1	255.255.255.0	无
Server0	Fa0/0	61.132.246.2	255.255.255.0	61.132.246.1

路由器 CR 的配置为：
```
//进入特权查看模式
enable
//进入配置模式
conf terminal
//设置路由器名称
hostname CR
//配置端口 Fastethernet0/0
ip address 202.102.216.2 255.255.255.0
ip nat outside
no shutdown
exit
//配置端口 fastethernet0/1.10
encapsulation dot1Q 10
ip address 192.168.0.1 255.255.255.128
ip nat inside
exit
//配置端口 fastethernet0/1.20
encapsulation dot1Q 20
ip address 192.168.0.129  255.255.255.128
ip nat inside
exit
//设置端口 fastethernet0/1
no shutdown
exit
//配置翻译后的出端口
ip nat inside source list 1 interface fastethernet0/0overload
//配置静态路由
ip route 0.0.0.0 0.0.0.0 fastethernet0/0
//配置想要翻译之前的地址
access-list 1 permit 192.168.0.0 0.0.0.255
```
路由器 ISP 的配置为：
```
//进入特权查看模式
enable
//进入配置模式
```

```
conf terminal
//设置路由器名称
hostname isp
//配置端口fastethernet0/0
interface fastethernet0/0
ip address 202.102.216.1 255.255255.0
no shutdown
exit
//设置端口fastethernet0/1
iP address 61.132.246.1 255.255.255.0
no shutdown
exit
```
交换机 SW1 的配置为：
```
//进入特权查看模式
enable
//进入配置模式
conf terminal
//设置交换机名称
hostname sw1
//创建vlan
vlan 10
exit
vlan 20
exit
//配置端口interface fastethernet0/2
switchport access vlan 10
switchport mode access
exit
//配置端口interface fastethernet0/3
switchport access vlan 20
switchport mode access
exit
//配置端口interface fastethernet0/1
switchport mode trunk
```
交换机 SW2 的配置为：
```
//进入特权查看模式
enable
//进入配置模式
conf terminal
//设置交换机名称
hostname sw2
//创建vlan
vlan 10
exit
vlan 20
exit
//配置端口interface fastethernet0/2
switchport access vlan 10
switchport mode access
exit
//配置端口interface fastethernet0/3
```

```
switchport access vlan 20
switchport mode access
exit
//配置端口 interface fastethernet0/1
switchport mode trunk
```

交换机 SW3 的配置为：

```
//进入特权查看模式
enable
//进入配置模式
conf terminal
//设置交换机名称
hostname sw3
//创建 vlan
vlan 10
exit
vlan 20
exit
//配置端口 interface fastethernet0/1
switchport mode trunk
exit
//配置端口 interface fastethernet0/2
switchport mode trunk
exit
//配置端口 interface fastethernet0/3
switchport mode trunk
exit
```

实验验证

从 PC0 ping 外网服务器 IP 地址正常，如图 35-2 所示。

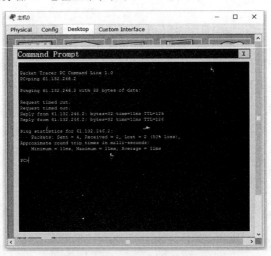

图 35-2　PC0 ping 外网服务器结果

从 PC1 ping 外网服务器 IP 地址正常，如图 35-3 所示。
从 PC2 ping 外网服务器 IP 地址正常，如图 35-4 所示。
从 PC3 ping 外网服务器 IP 地址正常，如图 35-5 所示。

实验 35　网吧网络设计（思科模拟器）

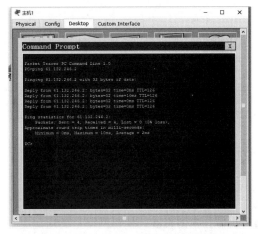

图 35-3　PC1 ping 外网服务器结果

图 35-4　PC2 ping 外网服务器结果

图 35-5　PC3 ping 外网服务器结果

拓扑设计中再增加一个防火墙应该怎样配置？

实验 36 校园网络设计（思科模拟器）

背景介绍

随着计算机、通信和多媒体技术的发展，使得网络上的应用更加丰富。同时因多媒体教育和管理等方面的要求，对校园网络也提出了进一步的要求。因此需要一个高速的、具有先进的、可扩展的校园计算机网络以适应当前网络技术发展的趋势并满足学校各方面的需要。信息技术的普及教育已经越来越受到人们关注。学校领导、广大师生们已经充分认识到这一点，学校未来的教育方法和手段，将构筑在教育信息化发展战略上，通过加大信息网络教育的投入，开展网络化教学，开展教育信息服务和远程教育服务等将成为未来建设的具体内容。

实验目的

掌握私有地址分配原则，能根据实际情况选择合适的子网掩码；掌握交换机的基本配置及划分 VLAN 的方法；掌握 OSPF 和 NAT 的基本原理。

实验步骤

1. 网络拓扑设计

网络拓扑设计如图 36-1 所示。

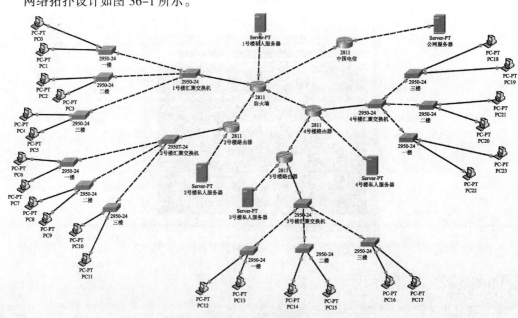

图 36-1　网络拓扑设计

实验所需要的设备为 24 台 PC 机，15 台交换机，5 台服务器，5 台路由器。其中有 4 台交换机作为汇聚交换机，剩下的 11 台交换机作为接入交换机；5 台服务器中的 1 台作为公网服务器，剩下 4 台作为私网服务器；5 台路由器中的 1 台作为防火墙，1 台作为 NAT 转换。作为防火墙的路由器因为端口不够需要加上两块板卡，除了作为中国电信 ISP 的路由器之外，其余的路由器需要加一块板卡，分别如图 36-2 和图 36-3 所示。

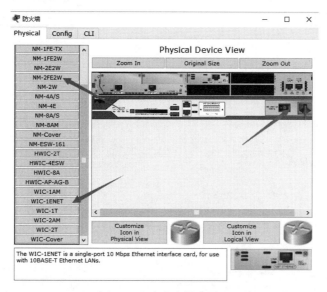

图 36-2　防火墙物理界面

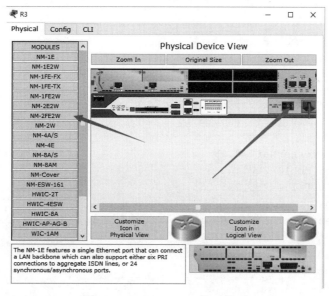

图 36-3　R1、R2、R3 物理界面

2．网络和服务器配置

各个设备及端口的 IP 地址规划如表 36-1 所示。

表 36-1 IP 地址规划

PC 终端	10.0.0.0-10.0.32.255
互联地址	192.168.0.0-192.168.0.255
私网服务器地址	100.0.0.0-100.0.0.255
防火墙出口地址	202.102.216.2/30
公网服务器地址	61.132.246.2/30

防火墙的主要配置为：

```
//进入特权查看模式
enable
//进入配置模式
conf terminal
//设定路由器名称
hostname CR
//配置 ABC 三个地址池
ip dhcp pool a
network 10.0.0.0 255.255.255.0
default-router 10.0.0.1
ip dhcp pool B
network 10.0.1.0 255.255.255.0
default-router 10.0.1.1
ip dhcp pool C
network 10.0.2.0 255.255.255.0
default-router 10.0.2.1
//配置 0/0 端口
interface fastethernet0/0
no shutdown
//配置子接口
interface fastethernet0/0.100
encapsulation dot1Q 100
ip address 10.0.0.1 255.255.255.0
ip nat inside
interface fastethernet0/0.200
encapsulation dot1Q 200
ip address 10.0.1.1 255.255.255.0
ip nat inside
interface fastethernet0/0.300
encapsulation dot1Q 300
ip address 10.0.2.1 255.255.255.0
ip nat inside
//配置 0/1 接口
interface fastethernet0/1
description to-ISP
ip address 202.102.216.1 255.255.255.252
ip nat outside
no shutdown
//配置 1/0 接口
interface fastethernet1/0
```

```
description to-china-union
ip address 100.0.0.1 255.255.255.252
no shutdown
//配置1/1接口
interface fastethernet1/1
description to-R1-fa0/1
ip address 192.168.0.2 255.255.255.252
ip nat inside
no shutdown
//配置0/3/0接口
interface ethernet0/3/0
description to-R2-fa0/1
ip address 192.168.0.6 255.255.255.252
ip nat inside
no shutdown
//配置ospf以及静态下发ospf
router ospf 100
network 10.0.0.0 0.0.255.255 area 0.0.0.0
network 192.168.0.0 0.0.0.255 area 0.0.0.0
network 100.0.0.0 0.0.0.255 area 0.0.0.0
default-information originate
//配置静态路由
ip route 0.0.0.0 0.0.0.0 fastethernet0/1
//配置nat
ip nat inside source list 30 interface FastEthernet0/1 overload
```

R2 的主要配置为：

```
//进入特权查看模式
enable
//进入配置模式
conf terminal
//设定路由器名称
hostname R2
//配置dhcp地址池
ip dhcp pool a
network 10.0.7.0 255.255.255.0
default-router 10.0.7.1
ip dhcp pool B
network 10.0.8.0 255.255.255.0
default-router 10.0.8.1
ip dhcp pool C
network 10.0.9.0 255.255.255.0
default-router 10.0.9.1
//配置0/0端口
no shutdown
//配置子接口
interface fastethernet0/0.100
encapsulation dot1Q 100
ip address 10.0.7.1 255.255.255.0
interface fastethernet0/0.200
```

```
   encapsulation dot1Q 200
   ip address 10.0.8.1 255.255.255.0
   interface fastethernet0/0.300
   encapsulation dot1Q 300
   ip address 10.0.9.1 255.255.255.0
//配置 0/1 端口
   interface fastethernet0/1
   description to-R1-fa1/1
   ip address 192.168.0.5 255.255.255.252
   ipv6 ospf cost 1
//配置 1/1 端口
   interface fastethernet1/1
   description to-R3-fa0/1
   ip address 192.168.0.10 255.255.255.252
//配置 1/0 端口
   interface fastethernet1/0
   description to-server-2
   ip address 100.0.0.9 255.255.255.252
//配置 ospf
   router ospf 100
   network 10.0.0.0 0.0.255.255 area 0.0.0.0
   network 192.168.0.0 0.0.0.255 area 0.0.0.0
   network 100.0.0.0 0.0.0.255 area 0.0.0.0
```

注意：R1、R3 的配置方法与路由器 R2 的配置方法相同，不再一一列举。

ISP 的主要配置为：

```
//进入特权查看模式
enable
//进入配置模式
conf terminal
//设定路由器名称
hostname isp
//配置 0/0 端口
interface fastethernet0/0
ip address 202.102.216.2 255.255.255.252
no shutdown
//配置 0/1 端口
interface fastethernet0/1
ip address 61.132.246.1 255.255.255.252
no shutdown
```

汇聚交换机的主要配置为：

```
//进入特权查看模式
enable
//进入配置模式
conf terminal
//设定路由器名称
hostname d-switch
//配置 0/1 端口
interface fastethernet0/1
description to-1f-fa0/1
```

```
switchport mode trunk
//配置0/2端口
interface fastethernet0/2
description to-2f-fa0/1
switchport mode trunk
//配置0/3端口
interface fastethernet0/3
description to-3f-fa0/1
switchport mode trunk
//配置0/4端口
interface fastethernet0/4
description to-nat-fa0/0
switchport mode trunk
```
注意：另外3台汇聚交换机配置类似，不再一一列举。

一楼的接入交换机主要配置为：
```
//配置0/1端口
interface fastethernet0/1
switchport mode trunk
//配置0/2端口
interface fastethernet0/2
switchport access vlan 100
//配置0/3端口
interface fastethernet0/3
switchport access vlan 100
```

二楼的接入交换机主要配置为：
```
//配置0/1端口
interface fastethernet0/1
switchport mode trunk
//配置0/2端口
interface fastethernet0/2
switchport access vlan 200
//配置0/3端口
interface fastethernet0/3
switchport access vlan 200
```

三楼的接入交换机主要配置为：
```
//配置0/1端口
interface fastethernet0/1
switchport mode trunk
//配置0/2端口
interface fastethernet0/2
switchport access vlan 300
//配置0/3端口
interface fastethernet0/3
switchport access vlan 300
```

实验验证

从一楼的PC中选出一台ping公网服务器IP地址正常，如图36-4所示。

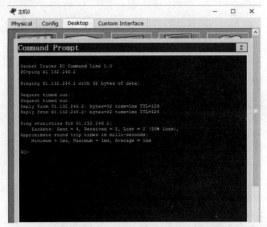

图 36-4　PC0 ping 公网地址实验结果

从二楼的 PC 中选出一台 ping 公网服务器 IP 地址正常，如图 36-5 所示。

图 36-5　PC8 ping 公网地址实验结果

从三楼的 PC 中选出一台 ping 公网服务器 IP 地址正常，如图 36-6 所示。

图 36-6　PC18 ping 公网地址实验结果

假如还有一个网络拓扑与此网络环境类似，应该如何配置静态路由使得该局域网与另外的局域网相连？

思科模拟器

1. 思科模拟器设备介绍（版本号：cisco packet tracer 6.1）

本书采取的 Cisco Packet Tracer 实验平台软件 是由 Cisco 公司发布的一个辅助学习工具，为学习思科网络课程的初学者去设计、配置、排除网络故障提供了网络模拟环境。用户可以在软件的图形用户界面上直接使用拖动方法建立网络拓扑，并可提供数据包在网络中行进的详细处理过程，观察网络实时运行情况。可以学习 IOS 的配置、锻炼故障排查能力。Packet Tracer 是一个功能强大的网络仿真程序，允许学生实验与网络行为，问"如果"的问题。随着网络技术学院的全面的学习经验的一个组成部分，包示踪提供的仿真、可视化、编辑、评估和协作能力，有利于教学和复杂的技术概念的学习。Packet Tracer 补充物理设备在课堂上允许学生用的设备，一个几乎无限数量的创建网络鼓励实践、发现和故障排除。基于仿真的学习环境，帮助学生发展如决策第二十一世纪技能，创造性和批判性思维，解决问题。Packet Tracer 补充的网络学院的课程，使教师易教，表现出复杂的技术概念和网络系统的设计。

2. 思科模拟器界面介绍

（1）常用工具栏：自上而下提供包括选择、移动、备注、删除、查看、绘制多边形。

（2）调整大小、添加简单或复杂数据包等常用工具。

（3）空白部分为工作区，可以在工作区创建网络拓扑结构，以及查看模拟网络中的数据。

（4）逻辑和物理工作区转换栏：逻辑工作区为主要工作区，用户在此工作区完成网络设备的逻辑连接和配置。物理工作区为模拟真实情况有城市、建筑物、工作间等直观图，用户对此进行相关配置。

（5）实时/模拟模式转换栏：实时模式为默认模式，模拟模式可模拟数据包的传输过程，更好查看整个网络拓扑图。

（6）网络设备库：给用户提供不同类型、不同型号的网络设备，包括路由器、交换机、集线器等。

（7）用户数据包窗格：用户可通过窗格管理数据包，添加、删除、查看数据包信息。

（8）基本操作，可把设备拖动到工作区，单击"连接线→设备→相连接→可选端口"。

（9）常见的连接线：自动连接线、console 配置线、直通线、交叉线、光纤、电话线、同轴电缆、dce 串行线、dte 串行线（从左到右），如图 A-1 所示。

图 A-1 思科路由器连线标志

3. 设备之间连线（在模拟器上）

下面介绍思科模拟器的设备如何连线：

先取两台路由器、一台交换机、一台 PC，如图 A-2 所示。

下面开始连线找到形如闪电的标志并单击，所有类型的线就展示出来，如图 A-3 所示。

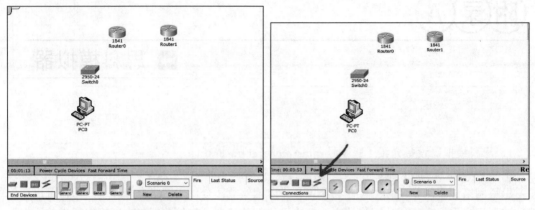

图 A-2　设备连线图 1　　　　　　图 A-3　设备连线图 2

上述所有设备均可用形如闪电标志的互联线，因为这种类型的线可以自动识别设备之间的接口类型，而随之改变，如图 A-4 所示。

所有设备之间线连接完毕后的结果如图 A-5 所示。

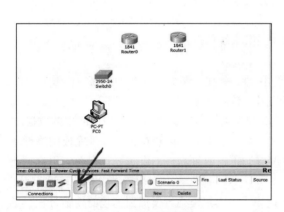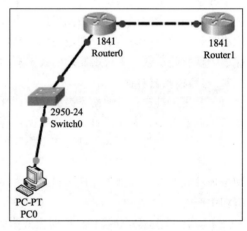

图 A-4　设备连线图 3　　　　　　图 A-5　设备连线图 4

附录 B 华为模拟器

1. 华为模拟器简介

eNSP（Enterprise Network Simulation Platform）是一款由华为提供的免费的、可扩展的、图形化操作的网络仿真工具平台，主要对企业网络路由器、交换机进行软件仿真，完美呈现真实设备实景，支持大型网络模拟，让广大用户有机会在没有真实设备的情况下能够模拟演练，学习网络技术。

华为模拟器（eNSP）下载和安装的过程如下：

（1）下载最新版本 VirtualBox（当前版本为 VirtualBox-4.3.26-98988）并安装，一直单击"下一步"按钮就可完成。

（2）下载华为工具 eNSP 最新版本及相应操作系统镜像如图 B-1 所示。

eNSP V100R002C00B360 Setup.exe
svrpbox.img

图 B-1 eNSP 新版本

（3）安装 eNSP，会弹出支持软件的相关工具软件，如图 B-2 所示。

图 B-2 选择安装其他程序

选择 svrpbox.img 文件，将该文件复制到"eNSP 安装路径\plugin\svrp\Database"即可。

（4）重启 Windows 系统。

（5）Windows 防火墙设置。

① 若是 Windows XP 版本，直接关闭防火墙。

② Windows 7 以上的版本，通过"控制面板→Windows 防火墙→允许程序或功能通过 Windows 防火墙"，如图 B-3 所示。

图 B-3　Windows 7 以上版本的相关设置

选择 eNSP 相关的复选框，如图 B-4 所示。

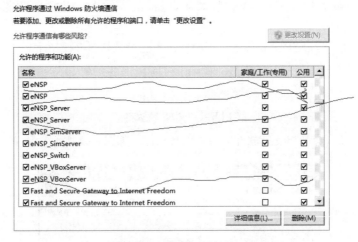

图 B-4　选择 eNSP 相关

（6）启动。

（7）根据系统及机器本身的物理结构的不同，可能会出现各种情况，只有根据实际情况，实际分析、解决。

2．eNSP 启动

（1）准备工作。为了更具真实性，要从主机直接访问网络设备。为此，应准备好 Microsoft Loopback 网卡。

① 如果是 Windows XP/2003 版本，按照"控制面板→添加硬件向导→是，我已经连接了此硬件→添加新的硬件设备→安装我手动从列表选择的硬件→网络适配器→Microsoft→Microsoft Loopback"。

② 如果是 Windows 7 版本，运行 hdwwiz，如图 B-5 ~ 图 B-7 所示。

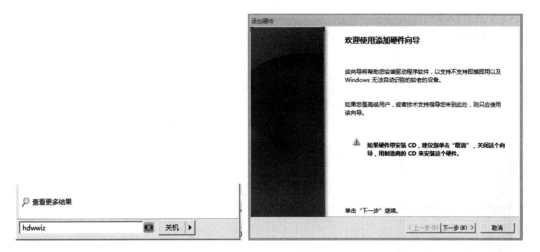

图 B-5　Windows 7 系统设置 1

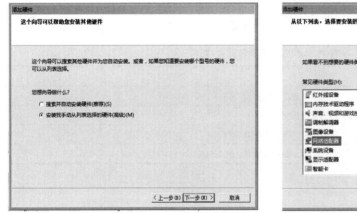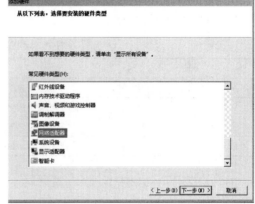

图 B-6　Windows 7 系统设置 2

图 B-7　Windows 7 系统设置 3

（2）将添加的 Microsoft Loopback 网卡 IP 地址设置为 192.168.1.101/24。

（3）搭建网络拓扑。

① 以管理员身份运行 eNSP，如图 B-8 所示。其运行界面如图 B-9 所示。eNSP 运行界面的各区域名称及简要描述如表 B-1 所示。

图 B-8　管理员身份运行 eNSP　　　　　　图 B-9　eNSP 运行界面

表 B-1　eNSP 区域名称及描述

序　号	区　域　名	简　要　描　述
1	快捷按钮	提供"新建"和"打开"拓扑的操作入口
2	样例	提供常用的拓扑案例
3	最近打开	显示最近已浏览的拓扑文件名称
4	学习	提供学习 eNSP 操作方法的入口

② 华为模拟器的设备连线

同样先取两台路由器、一台交换机、一台 PC 机，如图 B-10 所示。

　　R1　　　　　R2

　　LSW1

　　CLIENT1

图 B-10　设备连线图 1

下面开始连线，找到形如闪电的标志并单击，所有类型的线就展示出来，如图 B-11 所示。

上述所有设备均可用形如闪电标志的互联线，因为这种类型的线可以自动识别设备之间的接口类型，而随之改变，如图 B-12 所示。

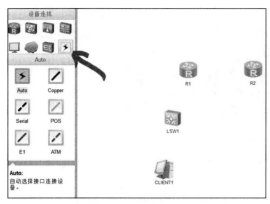
图 B-11 设备连线图 2

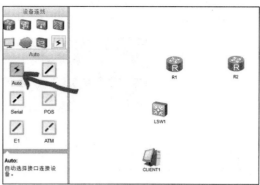
图 B-12 设备连线图 3

所有设备之间线连接完毕后的结果如图 B-13 所示。

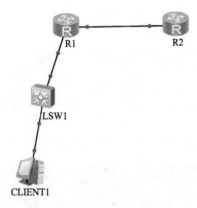

图 B-13 设备连线图 4

③ 搭建简单网络拓扑举例。

在 eNSP 模拟器中新建网络拓扑，并在路由器设备界面选取两台 AR 2220 路由器拖动到新建的拓扑图中；再从连接线界面中选取 AUTO 连接把两台路由器进行连接，此时连接默认选择 AR1 的 GE0/0/0 端口和 AR2 的 GE0/0/0 端口进行连接；最后再启动两台路由器，具体如图 B-14 所示。

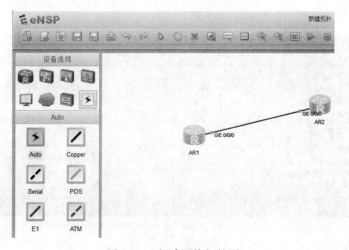

图 B-14 新建网络拓扑图

双击 AR1 路由器图标，进入路由器配置查看模式，再输入 sys 命令进入路由器配置模式，在配置模式下输入 sysname AR1 命令对此路由器的名称进行指定，具体命令格式如图 B-15 所示。

```
<R1>
<R1>sys
Enter system view, return user view with Ctrl+Z.
[R1]sysname AR1
[AR1]
```

图 B-15　进入配置模式和修改路由器名称

因为连接的是 AR1 的 GE0/0/0 端口，所以要对 AR1 的 GE0/0/0 端口进行配置。在配置模式下，输入 interface G0/0/0 命令进入端口配置模式；进入端口配置模式后，首先输入 undo shut 命令打开端口；然后用 ip add 100.0.0.1 255.255.255.252 的命令给端口添加 IP 地址和子网掩码，100.0.0.1 是 IP 地址，255.255.255.252 是子网掩码，因为是互联地址，所以采取了 30 位的子网掩码，30 位的子网掩码对应的子网包括 4 个 IP 地址，1 个地址为广播地址（100.0.0.3），一个地址为网络地址（100.0.0.0），恰好还有两个 IP 地址（100.0.0.1 和 100.0.0.2）分给两个路由器的端口，具体的具体命令格式如图 B-16 所示。

```
[AR1]int G0/0/0
[AR1-GigabitEthernet0/0/0]undo shutdown
Info: Interface GigabitEthernet0/0/0 is not shutdown.
[AR1-GigabitEthernet0/0/0]ip add 100.0.0.1 255.255.255.252
```

图 B-16　AR1 路由器端口配置

用同样的方法对 AR2 路由器进行配置，具体如图 B-17 所示。

```
<Huawei>sys
Enter system view, return user view with Ctrl+Z.
[Huawei]sysname AR2
[AR2]int G0/0/0
[AR2-GigabitEthernet0/0/0]undo shutdown
Info: Interface GigabitEthernet0/0/0 is not shutdown.
[AR2-GigabitEthernet0/0/0]ip add 100.0.0.2 255.255.255.252
```

图 B-17　AR2 路由器的配置

此时两台路由器配置完毕，随意选择一台路由器运行 ping 命令，ping 对端 IP 地址以检测两台路由器的通信是否正常（这里选择 AR2 路由器），具体如图 B-18 所示。

```
<AR2>ping 100.0.0.1
  PING 100.0.0.1: 56  data bytes, press CTRL_C to break
    Reply from 100.0.0.1: bytes=56 Sequence=1 ttl=255 time=450 ms
    Reply from 100.0.0.1: bytes=56 Sequence=2 ttl=255 time=70 ms
    Reply from 100.0.0.1: bytes=56 Sequence=3 ttl=255 time=30 ms
    Reply from 100.0.0.1: bytes=56 Sequence=4 ttl=255 time=20 ms
    Reply from 100.0.0.1: bytes=56 Sequence=5 ttl=255 time=30 ms

  --- 100.0.0.1 ping statistics ---
    5 packet(s) transmitted
    5 packet(s) received
    0.00% packet loss
    round-trip min/avg/max = 20/120/450 ms
```

图 B-18　ping 检测

ping 检测的结果正常，AR2 发出 5 的探测数据包都收到了正确应答。

H3C 设备简介

一、H3C 设备简介

H3C 设备主要包括：路由器、防火墙、二层交换机和三层交换机，分别介绍如下：

1. 路由器：H3C MSR36-20

软件版本号：Version 7.1.049。具体设备物理图如图 C-1 所示。

图 C-1　H3C MSR36-20

设备特性如下：

路由器类型：网络安全路由器。
端口结构：模块化。
广域网接口：3 个。
其他端口：2 个 USB 2.0 端口，1 个 CON/AUX 端口，1 个 CON 口。
扩展模块：4 个 SIC 插槽，2 个 DSIC，2 个 HMIM 插槽。
支持 VPN,QoS。
产品内存（默认/最大）：2GB/4GB。
FLASH：256 MB。
电源电压：AC/POE,100～240 V,50Hz/60Hz。
电源功率：125W。
产品尺寸：440 mm × 480 mm × 44.2 mm。
环境标准：环境温度 0～40℃。
其他特点：支持 3G。
IPv4 转发性能：5 Mpps。
IPv6 转发性能：4 Mpps。

2. 防火墙：H3C F100-C-SI

软件版本号：Version 5.20，具体物理设备如图 C-2 所示。

图 C-2　防火墙 H3C F100-C-SI

设备特性如下：
设备类型：企业防火墙。
网络端口：1 个配置口（CON）。
1 个 USB 接口。
2 个以太 WAN 接口。
4 个以太 LAN 接口。
支持 L2TP VPN，GRE VPN，IPSec/IKE，DVPN，SSL VPN。
电源：100～240 V。
产品尺寸：230 mm×160 mm×43.6 mm。
产品重量：1.8kg。
工作温度：0～45℃。
工作湿度：10%～95%，无冷凝。

3. 二层交换机：H3C s3100v2-26TP-EI

软件版本号：Version 5.20，具体物理设备如图 C-3 所示。

图 C-3　H3C s3100v2-26TP-EI

设备特性如下：
产品类型：快速以太网交换机。

应用层级：二层交换机。
传送速率：10/100 Mbit/s。
交换方式：存储转发。
背板带宽：32 Gbit/s。
包转发速率：6.6 Mbit/s。
端口结构：非模块化。
端口数量：28 个。
端口描述：24 个 10/100Base-TX 以太网端口，2 个 10/100/1000Base-T 以太网端口，2 个复用的 100/1000Base-X SFP 端口。
控制端口：1 个 console 口。
传输模式：全双工。
堆叠功能：可堆叠。
VLAN：支持基于端口的 VLAN（4K 个）。
　　　支持基于 MAC 的 VLAN。
　　　支持 GVRP。
　　　支持 VLAN VPN（QinQ）。
电源电压：AC 100-240V,50-60HZ。
电源功率：13W。
产品尺寸：360 mm × 160 mm × 43.6 mm。
工作温度：0 ~ 45℃。
工作湿度：10% ~ 90%（非凝露）。

4．三层交换机：H3C s3600v2-28TP-EI

软件版本号：Version 5.20，具体物理设备如图 C-4 所示。

图 C-4　H3C s3600v2-28TP-EI

设备特性如下：
产品类型：快速以太网交换机。
应用层级：3 层。
端口结构：非模块化。
端口数：28 个。
支持双工：是。
传送速率：10/100 Mbit/s。

交换方式：存储转发。

背板带宽：64 Gbit/s。

包转发率：9.6 Mbit/s。

端口描述：24 个 10/100 Mbit/s 端口，2 个 1000 Mbit/s SFP Combo 端口，2 个 1000 Mbit/s SFP 端口。

VLAN：支持基于端口的 VLAN(4K 个)。

支持基于协议的 VLAN。

支持基于 MAC 的 VLAN。

支持 Voice VLAN。

支持 Super VLAN。

支持 PVLAN。

支持 GVRP。

支持 VLAN VPN（QinQ），灵活 QinQ。

电源电压：AC 100–240V，50–60HZ。

尺寸：440 mm × 260 mm × 43.6 mm。

电源功率：31W。

工作温度：0 ~ 45℃。

工作湿度：10% ~ 90%（非凝露）。

二、H3C 网络设备登录

管理和配置网络设备有三种方法：第一是通过计算机终端连接网络设备的 console 口来管理；第二是通过 Telnet 协议远程管理网络设备；第三是通过 snmp 网络管理应用程序来管理网络设备。

本书在实验中是利用 console 口管理网络设备的，具体的步骤如下：

1. 安装 USB 串口线

（1）将 USB 串口线一端接计算机的 USB 接口，一端接交换机的 console 端口。注意这里 console 端口的一端不仅仅连接交换机，当然根据实验用到的设备也可以连接路由器设备、防火墙、三层交换机、AC、AP 等设备。这里以交换机为例进行介绍。

（2）安装 USB 串口线驱动程序。

① 首先找到计算机"控制面板"→"系统与安全"→"系统"→"设备管理器"，如图 C-5 所示。

② 找到设备管理器下的"其他设备"→"FT232R USB UART"，右击"FT232R USB UART"找到更新驱动程序软件下的"浏览计算机以查找驱动程序软件"，把桌面上的驱动程序软件文件夹添加上去并打开。此操作需要安装两次驱动软件程序（COM 口一次，USB 口一次），如图 C-6 所示。

③ 安装一次驱动软件程序后，原来的"FT232R USB UART"会变成"USB Serial Port"，如图 C-7 所示。

图 C-5　USB 串口线驱动安装 1

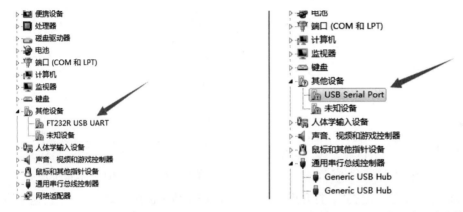

图 C-6　USB 串口线驱动安装 2　　　　　图 C-7　USB 串口线驱动安装 3

④ 安装第二次驱动程序软件文件，此时可以不用手动选择文件（不用选择"浏览计算机以查找驱动程序软件"），选择"自动搜索更新的驱动程序软件"即可，如图 C-8 所示。

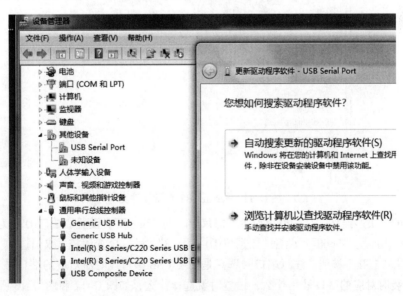

图 C-8　USB 串口线驱动安装 4

⑤ 驱动程序软件安装成功后，此时原来的"USB Serial Port"端口会显示为刚开始安装 USB 串口线时 USB 端口对应的 COM 口编号，如图 C-9 的"USB Serial Port（COM3）"所示。

安装完毕后，可以关闭所有程序，但是 console 线不能从计算机上拔下来，当然，和实体设备连接的 console 端口一端的插口，是可以拔出来换在其他实体设备上进行相应的配置的。

2. 设置交换机登录软件 SecureCRT

下面介绍交换机登录软件 SecureCRT 的具体使用方法，具体步骤如下：

（1）打开交换机登录软件 SecureCRT 的文件夹，找到可执行文件 SecureCRT 应用程序，如图 C-10 所示。

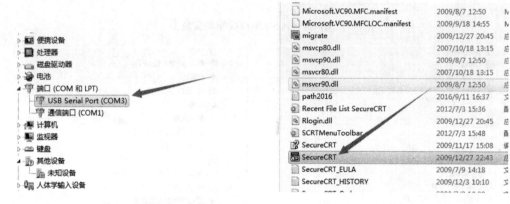

图 C-9　USB 串口线驱动安装 5　　　　图 C-10　SecureCRT 登录软件 1

（2）需要设置登录交换机方式，打开应用程序 SecureCRT 后，选择"文件"→"快速连接"命令，如图 C-11 所示。

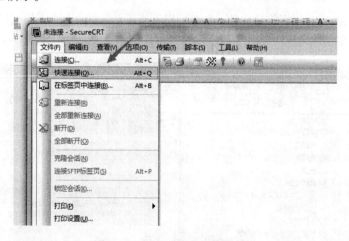

图 C-11　SecureCRT 登录软件 2

在弹出的"快速连接"对话框中设置"协议(P)"为"Serial"，设置"端口(O)"为"COM3"，设置"波特率(B)"为"9600"，设置"数据位(D)"为"8"，最后设置流控，取消选中"RTS/CTS"复选框，单击"连接"按钮。注意端口号选择是安装 USB 串口线时所对应的端口号（不一定是 COM3），安装时对应的端口号是什么，设置时就选择什么，如图 C-12 和图 C-13 所示。

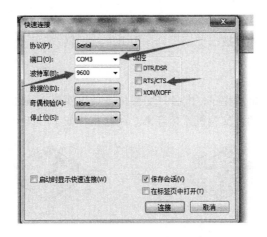

图 C-12　SecureCRT 登录软件 3

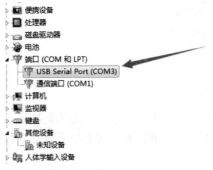

图 C-13　SecureCRT 软件需要的 COM 口编号

（3）连接后即可进入交换机登录软件管理界面，接下来进行其相应的实验配置命令即可，如图 C-14 所示。

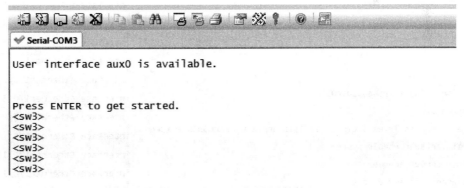

图 C-14　SecureCRT 登录软件 4

此时登录网络设备完毕，可以通过计算机终端对网络设备进行配置。

三、交换机端口的简易配置

1. 实验目的

掌握 H3C 交换机的端口的基本参数配置（双工、端口关闭）。

2. 实验设备

两台 H3C S3100 或 S3600 数据交换机，一根 RJ-45 的网线。

3. 实验步骤

（1）安装好串口线的驱动程序，准备好交换机登录软件 SecureCRT。

（2）用 RJ-45 网线把两台交换机连接起来。网络拓扑如图 C-15 所示。

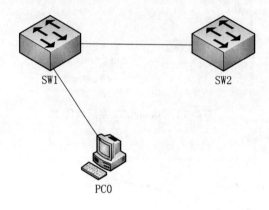

图 C-15　网络拓扑图

（3）在交换机运行配置文件中找到选取的端口。（配置文件是控制交换机如何工作的文件，其保存在交换机的硬盘上，交换机加电开机之后，其会加载到交换机的内存中控制交换机工作。）

① 在交换机查看模式下输入： dis cu，具体显示如图 C-16 所示。

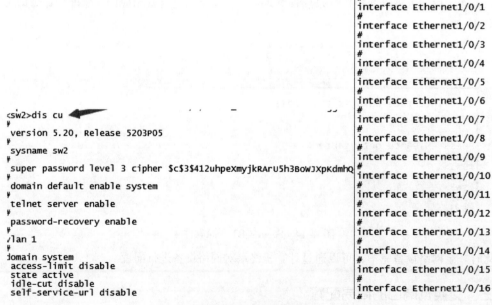

图 C-16　运行配置截图

交换机物理端口和运行配置中的端口是一对一对应的，以二层交换机 3100 为例，在配置里面就是 ethernet1/0/1，如果是 7 口就是 ethernet1/0/7。（交换机多采用多维数组的命名方式，

自左往右可以这样理解，板卡的位置/子卡在母板卡的位置/端口在子卡的位置）

② 查看一下所插入网线端口的状态。

```
<sw2>dis interface Ethernet 1/0/24
 Ethernet1/0/24 current state: UP（端口状态 UP 是正常、down 不正常，只要两端设备正常，网线正常，就是 up，跟协议无关）
 IP Packet Frame Type: PKTFMT_ETHNT_2, Hardware Address: 70f9-6da2-f0b5
 Description: Ethernet1/0/24 Interface
 Loopback is not set
 Media type is twisted pair（端口的传输介质是双绞线）
 Port hardware type is 100_BASE_T
 100Mbps-speed mode, full-duplex mode（端口是 100M、全双工工作模式）
 Link speed type is autonegotiation, link duplex type is autonegotiation
（端口速率和双工方式都是自适应的）
 Flow-control is not enabled
 The Maximum Frame Length is 2048
 Broadcast MAX-ratio: 100%
 Unicast MAX-ratio: 100%
 Multicast MAX-ratio: 100%
 PVID: 1
 Mdi type: auto
 Port link-type: access
 Tagged   VLAN ID : none
 Untagged VLAN ID : 1
 Port priority: 0
 Last clearing of counters: Never
 Peak value of input: 21 bytes/sec, at 2000-04-26 12:05:53
 Peak value of output: 19 bytes/sec, at 2000-04-26 12:05:53
 Last 300 seconds input:  0 packets/sec 14 bytes/sec 0%（最近 300 秒端口的进流量）
 Last 300 seconds output:  0 packets/sec 14 bytes/sec 0%（最近 300s 端口的出流量）
 Input (total):  170 packets, 46646 bytes
 2 unicasts, 2 broadcasts, 166 multicasts, 0 pause
```

③ 进入端口配置模式。

查看模式：<SW>

退出登录：quit

全局配置模式：在查看模式下输入 sys，就进入了全局配置模式。

退出全局配置：输入 quit，即可退出全局模式，并回到查看模式。显示如下：

```
<sw2>sys
[sw2]
[sw2]quit
<sw2>
```

端口配置模式：在全局配置模式下输入插入网线的端口号。

```
[sw2] interface Ethernet1/0/24
[sw2-Ethernet1/0/24]
```

退出端口模式：同样是敲入 quit，即可进入全局配置模式。

```
[sw2-Ethernet1/0/24]quit
[sw2]
```

④ 配置端口工作模式。

先配置一台交换机的双工工作方式为强制半双工：

```
sw2-Ethernet1/0/24]duplex half
```

怎样验证配置已经成功，操作如下：

验证状态（到查看状态下）：

```
[sw2-Ethernet1/0/24]dis this
#
interface Ethernet1/0/24
 duplex half
#
<sw2>dis int Ethernet 1/0/24
 Ethernet1/0/24 current state: UP
 IP Packet Frame Type: PKTFMT_ETHNT_2, Hardware Address: 70f9-6da2-f0b5
 Description: Ethernet1/0/24 Interface
 Loopback is not set
 Media type is twisted pair
 Port hardware type is  100_BASE_T
 100Mbps-speed mode, half-duplex mode(已经变成半双工了)
 Link speed type is autonegotiation, link duplex type is force link
```

怎样删除配置？

```
[sw2-Ethernet1/0/24]undo duplex
```

验证状态（到查看状态下）：

```
[sw2-Ethernet1/0/24]dis this
#
interface Ethernet1/0/24
#
```

⑤ 关闭端口操作。

```
[sw2]inter
[sw2]interface Ethernet 1/0/24
[sw2-Ethernet1/0/24]shutdown
[sw2-Ethernet1/0/24]
[sw2-Ethernet1/0/24]dis this
#
interface Ethernet1/0/24
 shutdown
#
return
[sw2-Ethernet1/0/24]quit
[sw2]quit
<sw2>dis int ether
<sw2>dis int Ethernet 1/0/24
 Ethernet1/0/24 current state: DOWN ( Administratively )（表示被管理 down，端口关电）
 IP Packet Frame Type: PKTFMT_ETHNT_2, Hardware Address: 70f9-6da2-f0b5
 Description: Ethernet1/0/24 Interface
 Loopback is not set
 Media type is twisted pair
 Port hardware type is  100_BASE_T
```

```
  Unknown-speed mode, unknown-duplex mode
  Link speed type is autonegotiation, link duplex type is autone
```

查看另一个交换机状态:
```
<sw3>dis interface Ethernet 1/0/24
 Ethernet1/0/24 current state: DOWN（收不到对端的电信号，端口 down）
 IP Packet Frame Type: PKTFMT_ETHNT_2, Hardware Address: 70f9-6da2-eb55
 Description: Ethernet1/0/24 Interface
```
⑥ 打开端口：[sw2-Ethernet1/0/24]undo shutdown。

此时端口打开。

端口简易配置命令介绍完毕，注意这里只是以交换机为例来介绍简易的端口配置命令，如果想了解设备的端口完整的配置命令，还需要查阅相关的资料。